RADICAL CONSTRUCTIVISM IN MATHEMATICS EDUCATION

Mathematics Education Library

VOLUME 7

The titles published in this series are listed at the end of this volume.

RADICAL CONSTRUCTIVISM IN MATHEMATICS EDUCATION

Edited by:
ERNST VON GLASERSFELD
University of Massachusetts

KLUWER ACADEMIC PUBLISHERS
DORDRECHT / BOSTON / LONDON

ISBN 978-90-481-4091-6 e-ISBN 978-0-306-47201-5

Published by Kluwer Academic Publishers,
P.O. Box 17, 3300 AA Dordrecht, The Netherlands.

Kluwer Academic Publishers incorporates
the publishing programmes of
D. Reidel, Martinus Nijhoff, Dr W. Junk and MTP Press.

Sold and distributed in the U.S.A. and Canada
by Kluwer Academic Publishers,
101 Philip Drive, Norwell, MA 02061, U.S.A.

In all other countries, sold and distributed
by Kluwer Academic Publishers Group,
P.O. Box 322, 3300 AH Dordrecht, The Netherlands.

Printed on acid-free paper

TABLE OF CONTENTS

ABOUT THE AUTHORS

While this volume was being prepared, the editor requested from the contributing authors a brief statement concerning their past intellectual itinerary and present ideas. These statements, adjusted in length, are here prefaced by indications of the authors' present academic positions.

Nicolas Balacheff *(Currently research director at the French National Center for Scientific Research (CNRS), president of the International Group for the Psychology of Mathematics Education (PME), and editor of the international journal "Recherches en Didactique des Mathematiques". He has recently joined the Laboratoire IRPEACS (CNRS), where he directs a research project on principles for the design of Intelligent Tutorial Systems).*

I have been involved in mathematics education since 1977. For the past ten years the main focus of my research was on how students learn mathematical proofs, with specific attention to how they deal with counter examples and refutations. I did most of this work as a member of the *Equipe de Didactique des Mathematiques et de l'Informatique* at the Joseph Fourier University in Grenoble. There I taught mathematics and mathematics education from 1972 to 1988. All along I have been concerned with the application of research results to practice and in particular to the practice of teacher education. The "intelligent tutorial systems" I am at present helping to design are dependent on the elaboration of a model of the active learner. Hence my profound interest in what and how students think.

Paul Cobb *(Ph.D. in Mathematics Education, University of Georgia, 1983. Currently directing research on mathematics instruction in the first three grades at Purdue University).*

Having initially had several years of intensive interaction with Les Steffe, Ernst von Glasersfeld, and John Richards on a project on young children's construction of arithmetical knowledge, I was much influenced by radical constructivism as a theory of knowing and as an orienting framework for educational issues. However, since 1983, I have been troubled by sociological and cultural phenomena which seemed to fall outside my admittedly immature grasp of constructivism. This feeling was further exacerbated by time spent in a classroom during a year-long teaching experiment. I have come to the view that

E. von Glasersfeld (ed.), Radical Constructivism in Mathematics Education, vii–xi.

it is impossible to explain children's mathematical learning unless psychological analyses are coordinated with anthropological analyses of classroom life. In this regard, I have been much influenced by a variety of theorists whose basic epistemology is compatible with constructivism (e.g. Goodman, Bloomer, Schutz, Mehan, Bauersfeld, Voigt, Barnes, Geertz, and Rorty). I now view classrooms as sites for action research in which to explore salient questions by having conversation-like interactions with the providers of one's "data".

Jere Confrey *(Ph.D. in Mathematics Education, Cornell University. Founder of Summer Math Program, Mount Holyoke College, Massachusetts. Currently directing research projects on exponential functions and multi-representational software at Cornell University.)*

I have come to constructivism through my interest in philosophical issues concerning knowledge and philosophy of science. As feminist, mother, and professor I had little difficulty in accepting the experience of multiple realities and I discovered my commitment to work against the silencing of student voices that is prevalent in our traditional ways of schooling. I met Ernst von Glasersfeld some ten years ago and he has continued to encourage me to examine my own understanding of cognition and mathematics. He, Les Steffe, and Paul Cobb have contributed to the development of my ideas about constructivism. My current work leads me towards the design of technological tools that promote students' creativity rather than stifling it.

James Kaput *(Professor of Mathematics, Southeastern Massachusetts University; Coordinator of Mathematics Projects, Educational Technology Center, Harvard Graduate School of Education; Research Associate, National Center for Research in Mathematical Sciences Education.)*

My intellectual odyssey began when, having done my doctoral work as a mathematician in category theory, Bob Davis encouraged me in the early 1970's to turn my attention to mathematics education. I became interested in finding better ways for undergraduates to learn mathematics and in preparing myself to help other teachers. Along the way it became clear to me that psychology, philosophy, mathematics, linguistics, anthropology, information science, and almost everything else I study and appropriate, are all part of the same enterprise. From my individual perspective, they all are branches of the science of re-presentation. Now I am especially intrigued by the fact that the act of building computer-based learning environments – and trying to make them work – forces me simultaneously to apply everything I have learned *and* to make my beliefs explicit and concrete. That all this is not an idle academic exercise, is regularly made plain by my three kids who are struggling through traditional school math, traditionally re-presented.

Clifford Konold *(Research Associate at the Scientific Reasoning Research Institute, University of Massachusetts. Currently directing the NSF-funded* *project "A Computer-Based Curriculum for Probability and Statistics".)*

As a graduate student in psychology I was interested in studying the effects of instruction on (as I would have said then) "the structure of concepts in long-term memory". Probability and statistics offered a convenient research domain, both because I was teaching the topic and because it had a more definite, or agreed upon, structure than domains such as "Educational Psychology". I soon became dissatisfied with the techniques that were then being used in psychology to characterize memory structure and found myself struggling to adapt Piaget's clinical approach to my interests. I fell in with a group of like-minded individuals, got divorced, parted company with the religious group in which I had been raised, and finally convinced myself that it was OK to pay someone to do auto repairs that I was theoretically capable of doing myself. As a result of these changes, I now spend much less time on my back and my knees – all this because I wanted a different way to think about structure.

Jack Lochhead *(Graduate work in physics and Doctorate in Educational Research and Statistics, University of Massachusetts, 1973. Currently Director of Scientific Reasoning Research Institute, Director Basic Math Program, and Assistant Dean for Natural Science and Mathematics, University of Massachusetts.)*

My interest in scientific and mathematical thinking stems from my own struggle to understand some of the apparently simple concepts in those fields. I believe that once we develop a clearer picture of scientific thinking and understanding, we may be able to design more efficient approaches to instruction. If we find out more about how both students and physicists come to have the concepts with which they operate, we should have a better chance to eliminate the discrepancies and bridge the gaps that the present educational practices seem to leave untouched. Whenever I encounter evidence of serious inconsistencies in current models of thinking, I am inspired by the potential of an advance in education.

John Richards *(Ph.D. in Philosophy, State University of New York at Buffalo. Currently Software Development Manager at BBN Labs, Cambridge, Massachusetts.)*

My commitment to a constructivist view of knowledge grew out of research in the history of science and the foundations of mathematics. While teaching philosophy at the University of Georgia, I became convinced that a sound theory of knowledge would have some revolutionary practical implications if it were applied to teaching in schools and general learning situations. I am currently exploring the development and use of intelligent software tools that provide structured environments for learning. An approach we have called "Inquiry by Design" uses the computer environment to guide the students' active exploration in mathematics and especially in algebra.

Philip H. Steedman *(Ph.D. in Philosophy, University of Illinois. Graduate work at University of Canterbury. Currently teaching Philosophy of Education at the University of Cincinnati.)*

During my post-doctoral work as Research Associate at the University of California at Berkeley and at the Institute of Advanced Studies, Princeton, I became interested in the implications of philosophy of science and hermeneutics and their application to the production of knowledge and distribution within a public school system. Several years of experience as a professional music critic gave me the opportunity to see parallels between art and science and the performance of teachers.

Leslie P. Steffe *(Appointed in 1967 to the faculty of Mathematics Education at the University of Georgia. He has directed a succession of research projects involving microanalysis of young children's number concept and mathematical thinking and is currently a Research Professor and Fellow of the Institute of Behavioral Research at the University of Georgia.)*

Through interdisciplinary work on understanding the operations that children use to generate their mathematical knowledge, I came to the conviction that children's generative power in mathematics is almost totally uncharted. I believe that, rather than expecting children to learn how the teacher thinks, mathematics teachers must teach in order to learn how children think. Teachers are in the best position to establish a phenomenology of children's ways and means of operating. Once such a phenomenology has been established, teachers can begin to teach in harmony with children's approach to mathematics. I believe that it is absolutely essential for teachers to create deep and sustained communication in the mathematics classroom, and thus create an environment that fosters the students' active development of mathematical knowledge.

Bob Underhill *(Graduate studies at Purdue University; Ed.D. in Mathematics Education, Michigan State University, 1968. He has taught at Ohio University, the University of Houston, and Kansas State University. Currently he is Professor of Mathematics Education at Virginia Polytechnic Institute and State University in Blacksburg, Virginia.)*

My professional journey led me to a major encounter with Piaget's work in the late 1960's. I picked and chose ideas for which I was ready, mainly developmental aspects of learning; I coupled these ideas with Bruner's enactive, iconic, and symbolic knowing. For several years I used these concepts in a competence-based orientation in teacher education (CBTE). Over the years I became disenchanted because I could not overcome the *can do* with a *will do* in CBTE. This led me in recent years to an interest in beliefs and the theory of constructivism. I am still attempting to translate my own beliefs into teacher education, since I believe that the practices in a teacher education program must mirror practices in schools.

Jan van den Brink *(Senior member of the Research Group on Mathematics Education and the Educational Computer Center of the State University at Utrecht, The Netherlands.)*

I started out as a teacher in primary education, became a mathematician, and was involved, from its beginning in 1971, in the I.O.W.O. (Institute for Development of Mathematics Instruction), which became famous for its view on mathematics as a human activity and curriculum development as a classroom activity. The main thrust of my educational efforts was to develop a curriculum in which notions and ideas that come more or less naturally to children can be fitted into mathematical structures.

I believe that not only does the teacher observe the children in his class, but the children also observe the teacher. Like the children, we, as researchers in education, can learn much from good teachers. We can be inspired to find new research methods and techniques by closely watching an effective teacher's ways and means of teaching. But to formulate our observations, we need a viable theory of what we consider to be the knowledge that is to be taught.

Terry Wood *(Ph.D. in Elementary Education, Michigan State University; currently Assistant Professor of Mathematics Education, at Purdue University.)*

For a long time I have known that the question of the particular meaning one gives to experiences has been a critical aspect of my own thinking and learning. Now, in trying to understand the learning of young children, I still believe that experience and personal meaning are essential, but I have come to realize the importance of social interaction. This perspective was initially engendered by my experiences as a teacher, in graduate studies, with my own children, and more recently in work with Paul Cobb.

Erna Yackel *(Ph.D. in Mathematics Education, Purdue University, 1984; currently Assistant Professor of Mathematics Education, Purdue University, Calumet.)*

My educational convictions derive from a variety of experiences: teaching mathematics and statistics at the university level, developing and conducting a program to help students overcome mathematics anxiety, and curriculum development for both elementary and secondary schools. The use of small group program solving (SGPS), including total group discussions and deliberate attempts to influence individuals' attitudes toward mathematics, were so successful in the Overcoming Math Anxiety Program that I subsequently used SGPS in my regular university courses as well as in curriculum development for high school mathematics. The psychological research with Paul Cobb has helped me to find a theoretical basis compatible with the intuitions that guided my former work.

INTRODUCTION

> Mathematics is the science of *acts without things* – and through this, of things one can define by acts.
>
> Paul Valéry[1]

The essays collected in this volume form a mosaik of theory, research, and practice directed at the task of spreading mathematical knowledge. They address questions raised by the recurrent observation that, all too frequently, the present ways and means of teaching mathematics generate in the student a lasting aversion against numbers, rather than an understanding of the useful and sometimes enchanting things one can do with them.

Parents, teachers, and researchers in the field of education are well aware of this dismal situation, but their views about what causes the wide-spread failure and what steps should be taken to correct it have so far not come anywhere near a practicable consensus.

The authors of the chapters in this book have all had extensive experience in teaching as well as in educational research. They approach the problems they have isolated from their own individual perspectives. Yet, they share both an overall goal and a specific fundamental conviction that characterized the efforts about which they write here. The common goal is to find a better way to teach mathematics. The common conviction is that knowledge cannot simply be transferred ready-made from parent to child or from teacher to student but has to be actively built up by each learner in his or her own mind.

My purpose in compiling this book, and the purpose of the authors who have contributed its chapters, is not to make yet another move in the theoretical philosophical debate about constructivism. The book is intended to clarify how the didactic attitude changes when the constructivist theory of knowing is put into practice, and what results have been attained in doing so. Rather than forecasts of what might be achieved in the future, the papers assembled here report on experiments and implementations that have actually been carried out.

The individual pieces speak for themselves and require no amplifications. They cover a wide range of educational efforts: Elementary school (Cobb, Steffe, van den Brink), high school (Balacheff, Richards), undergraduate instruction (Confrey, Kaput, Konold, Lochhead), and teacher preparation (Underhill). Consequently they deal with different areas of mathematics, and it

E. von Glasersfeld (ed.), Radical Constructivism in Mathematics Education, xiii–xx.

may be useful to anticipate and spell out here some of the principles that underlie the constructivist approach common to all the contributions.

* * *

Language frequently creates the illusion that ideas, concepts, and even whole chunks of knowledge are transported from a speaker to a listener. This illusion is extraordinarily powerful because it springs from the belief that the meaning of words and phrases is fixed somewhere outside the users of the language. Perhaps the best way to dismantle the illusion is to remember or reconstruct how one came to form the meanings of words and phrases when one was acquiring language in the first place. Clearly it could only be done by associating bits of language one heard with chunks of one's own experience – and no one's experience is ever exactly the same as another person's. Thus, whatever another says or writes, you cannot but put your own *subjective* meanings into the words and phrases you hear. Given that we live in a community of other language users, our subjective meanings tend, of course, to become *intersubjective*, because we learn to modify and adapt them so that they fit the situations in which we interact with others. In this way we manage to achieve a great deal of compatibility. But to prove compatible, individual meanings do not have to be identical. Indeed, throughout our lives we now and then discover that the meaning we have associated with a certain word is not yet quite compatible with the use others make of that word.[2] This may serve to remind us – especially when we act as teachers – that new concepts and new knowledge cannot simply be passed to another person by talk, because each must abstract meanings, concepts, and knowledge from his or her own experience.

This does not mean that language cannot be used to *orient* (Maturana, 1980) students towards certain experiences and certain mental activities such as abstracting; but it does mean that we can never rely on language to "convey" knowledge as though it were something like food that can be handed from one to another.

* * *

The notion that knowledge is the result of a learner's activity rather than of the passive reception of information or instruction, goes back to Socrates and is today embraced by all who call themselves "constructivists". However, the authors whose work is collected here, constitute the radical wing of the constructivist front. They have taken seriously the revolutionary attitude pioneered in the 1930s by Jean Piaget, the Swiss founder of cognitive psychology. This attitude is characterized by the deliberate redefinition of the concept of knowledge as an *adaptive function*. In simple words, this means that the results of our cognitive efforts have the purpose of helping us to cope in the world of our experience, rather than the traditional goal of furnishing an "objective" representation of a world as it might "exist" apart from us and our experience.[3] This attitude has much in common with the pragmatist ideas proposed by William James and John Dewey at the beginning of this century. It is *radical* because it breaks with the traditional theory of knowledge and it has profound consequences for parents, teachers, and researchers whose objective is to generate particular ways of acting and thinking in children and students. Steedman's essay grounds the changed attitude in the philosophy of science and the other contributors illustrate some of the consequences for the practice of education.

* * *

When teachers look at a student in their class, they see that student in a complex environment. There is the physical environment of the desk (or whatever piece of furniture the student occupies), the classroom, and the land- or townscape around the school building, part of which might be visible through the window; there is the social environment of the classmates, the teachers themselves, and the student's family, which may to some extent be known to the particular teachers, and there is also the didactic environment of the school books, the educational props, the curriculum, and, more important but less perceivable, the teachers' views and beliefs as to how education should be achieved.

From the naive commonsense perspective, the elements that form this complex environment belong to a *real* world of unquestionable objects, as *real* as the student, and these objects have an existence of their own, independent not only of the student but also of the teacher.

Radical Constructivism is a theory of *knowing* which, for reasons that had nothing to do with teaching mathematics or education, does not accept this commonsense perspective.[4] Instead, it takes seriously the no less venerable suggestion that what a teacher sees when he or she looks at a student is necessarily part of the particular teacher's experience and, as such, the result of the particular teacher's ways and means of perceiving and conceptualizing what is being preceived. Thus, the teacher's view of the student, the classroom and its furniture, the surroundings of the school house, the classmates, the student's family, the school books and other educational material, the curriculum, and, needless to say, the teacher's views about education, are and cannot be anything but the particular teacher's own experiential world.

Superficial or emotionally distracted readers of the constructivist literature have frequently interpreted this stance as a denial of "reality".[5] If everyone had a different experiential world, they tend to argue, we could not agree on anything and, above all, we could not communicate. There is not much wrong with that argument, but the fact that we *do* agree on certain things and that we *can* communicate does not prove that what we experience has *objective reality* in itself. If two people or even a whole society of people look through distorting lenses and agree on what they see, this does not make what they see any more *real* – it merely means that on the basis of such agreements they can build up a consensus in certain areas of their subjective experiential worlds. Such areas of relative agreement are called "consensual domains", and one of the oldest in the Western world is the consensual domain of numbers. The certainty of mathematical "facts" springs from mathematicians' observance of agreed-on ways of operating, not from the nature of an objective universe. It is a shared belief of our authors that establishing a consensual domain that comprises the instructor as well as the learners is a prerequisite of teaching.

* * *

From an educator's point of view, one of the most important features of radical constructivism is the sharp distinction it draws between teaching and training. The first aims at generating *understanding*, the second at *competent performance*. The fifty years of behaviorist hegemony in American psychology all but succeeded in eliminating that distinction. The sophisticated techniques developed by Professor Skinner and his followers have shown that in many tasks almost flawless performance can be achieved by the methodical management of stimuli and reinforcement. Yet, if our purpose is to spread mathematical knowledge, these techniques are inappropriate because the "competence" they engender has been deliberately cut loose from mental operations and understanding.[6] To know mathematics is to know how and why one operates in specific ways and not in others, how and why the results one obtains are derived from the operations one carries out.

The focus on understanding leads to several other considerations. As long as students merely strive to find answers that are "right" because they satisfy the teacher, the students may learn to get through class more or less painlessly, but they will not learn mathematics. The early dialogs in John Richards' chapter are a blatant example of this aberration. When that spurious goal has been eliminated, another point comes to the fore: Students who are trying to solve a problem, rather than please the teacher, do not answer randomly. What they produce makes sense to *them* at the time, no matter how little sense it might make to the more advanced mathematician. And it is this sense that the teacher must try to understand if he wants to find ways and means of modifying the students' thinking. Several of our authors emphasize this aspect of the teacher's task. Steffe calls it *constructing a model of the student's concepts and operations* and we all consider it indispensable as a working hypothesis for the teacher who intends to guide the student towards the kind of mathematics to be learned.

However, the models we construct of other people's ideas and mental operating, regardless of whether we are concerned with their politics or their mathematics, are necessarily hypothetical because we have no direct access to what goes on inside other people's heads. Nevertheless we can test the *viability* of our models of students' thinking by creating situations in which these ways of thinking would be likely to produce certain observable results. This is no different in principle from physicists who set up experiments to test their models of unobservable sub-atomic particles. The traditional psychologists' objection to mental explanations, thus, is nothing but the remnant of a superannuated early 19th-century view of "science".

Indeed, if we are serious in maintaining that learning aims at *understanding*, we should avoid the term "mathematical literacy" that has recently become a cliché of mathematics education. Research on reading was for a long time hampered by the shallow notion that literacy was simply a matter of recognizing written or printed letters and being able to reproduce words vocally. Only when comprehension on the conceptual level was taken into account, did the research in that area make some progress. Similarly, to recognize a sequence of spoken or written mathematical symbols as the stimulus for certain conventional operations may be a prerequisite of conceptual understanding, but it is no more an indication of comprehension than the ability vocally to produce the sentence "time and space form a continuum" would indicate an *understanding* of Einstein's theory of relativity.

The models we construct of other people's thought processes can, of course, never be anything but *our* constructs built of components that are accessible to *us*. But this inescapable limitation does not mean that we cannot use our cognitive ability to construct a view and an interpretation of the given situation that turns out to be more compatible with what we perceive of the other's actions and reactions. Not surprisingly, this effort may lead the teacher to a new and fuller understanding of the subject area (Confrey, Underhill). Yet as long as teachers cling to the traditional conviction that the solutions to mathematical problems are not only obvious but also unique, and that the failure to *see* them is due either to stupidity or a lack of application, they have little if any incentive to investigate whatever sense the students might make of a problem and their tentative solutions.

As an important corollary, the teacher's attempts to understand the individual student's approach to a "problem" generate a climate of positive social interaction. Genuine interest in how they think shows the students, better than any verbal affirmation, that they are being taken seriously; and this, in turn, enhances their courage to try and openly discuss new paths of whose outcome they are still uncertain. There is no more efficient way to generate the kind of *reflection* that is necessary for conceptual advancement (Cobb, Confrey, Kaput, Richards, Steffe, Underhill, van den Brink).

* * *

Although Piaget, for more than five decades, maintained that all "operative" knowledge (i.e., the know-how concerning mental operations) was the result of reflection, psychologists who claimed to be strict empiricists took great pains to avoid any mention of the term. This is odd, because one of the few things

Locke, Berkeley, and Hume, the founders of Empiricism, agreed on, was that all knowledge comes either from the senses or from reflection, which Locke had defined as the mind's ability to ponder its own operations. (To avoid embarrassment about not having acknowledged the role of reflection for so long, the psychological establishment is now rediscovering at least parts of it under the name of "metacognition".)

From the constructivist point of view, there can be no doubt that reflective ability is a major source of knowledge on all levels of mathematics. That is the reason why nearly all the contributors to this volume consider it important that students be led to talk about their thoughts, to each other, to the teacher, or to both. To verbalize what one is doing ensures that one is examining it. And it is precisely during such examination of mental operating that insufficiencies, contradictions, or irrelevancies are likely to be spotted (Balacheff, Cobb, Confrey, Lochhead, Kaput, Konold, Richards, van den Brink).

* * *

A thinking subject has no reason to change his or her way of thinking as long as there is no awareness of failure. But there are many shades and forms of failure, and the one most common in conventional schools – when the teacher declares that the student is "wrong" – is the least effective in bringing about a change in the student's way of thinking. Far more powerful is the failure to gain the agreement of peers who are struggling with or have solved the same problem. As Piaget reiterated many times, after infancy the most frequent cause of *accommodation* (change in a way of operating or acting) arises in social interaction when the individual's ways and means turn out to be in some sense insufficient in comparison to the ways and means of others (Piaget, 1967; cf. also Rowell, 1989).

Analogously, a thinking subject has no occasion to feel the intellectual satisfaction of having solved a problem, if the solution did not result from his or her own management of concepts and operations but was supplied from outside. Here again the behaviorist notion of social approval as the prime reinforcement has helped to disorient schooling practice. It is not that a teacher's approval and pat on the head have no effect on the student, but the effect is to strengthen the student's inclination to please the teacher rather than to build up *understanding* of the conceptual *area* in which the task was situated. Thus students are prevented from experiencing the rewarding elation that follows upon having found *one's own way and recognizing it as a good way*. If students are not oriented or led towards autonomous intellectual satisfaction, we have no right to blame them for their lack of proper *motivation*. The motivation to please superiors without understanding why they demand what they demand, may be required in an army – in an institution that purports to serve the propagation of knowledge, it is out of place.

* * *

The empirical investigations and the applications of constructivist ideas in teaching and learning situations were carried out and are here reported by highly original researchers, each of whom has developed an individual style. The theoretical underpinnings, however, are largely compatible if not the same. Radical Constructivism, I want to emphasize, is a theory of *active knowing*, rather than a traditional theory of knowledge or epistemology. From this standpoint, as Piaget maintained fifty years ago, *knowledge serves to organize experience*, not to depict or represent an experiencer-independent reality.

One of the main sources of the constructivist theory was an analysis of linguistic communication that exposed the subjective roots of the *meanings* in terms of which each individual interprets the language heard or read. For educators, this reinforces the intuition that the student's experiential reality is not the same as the teacher's, and it makes explicit the fact that, because language does not "transport" knowledge, it must and can be used to *orient* a student's own conceptual construction.

The claim that thinking individuals create for themselves individual realities does not imply, as critics of constructivism tend to assert, the denial of an external reality; it claims no more than it literally says, namely that individuals organize their experience in their own subjective ways.

The constructivist theory, consequently, clearly distinguishes *training* from *teaching*. The former may lead to the replication of a behavioral response, the latter aims at generating *autonomous conceptual understanding*.

Teachers have a better chance to modify a student's conceptual structures and understanding if their interventions are informed by a hypothetical model of the student's present ideas. By reinforcements or threats, they may get the student to repeat anything they say, but understanding cannot be forced in this way. Moreover, the mere acceptance of the solution of a problem that is forced on them, is not likely to give students a taste of the kind of intellectual satisfaction that would generate a genuine motivation to understand more and to delve further into the problem area.

In contrast, leading students to discuss *their* view of a problem and their own tentative approaches, raises their self-confidence and provides opportunities for them to reflect and to devise new and perhaps more viable conceptual strategies. This is one among several reasons why all the contributions to this volume stress the importance of the social interaction and the social climate the teacher establishes in the classroom.

As a constructivist, I am fully aware of the fact that the introductory notes I have presented here cannot determine the interpretation of the essays that follow. But they may orient the reader to look for some of the things that provided the original impetus for generating this volume.

E.v.G.

July 1989 Amherst, Massachusetts

NOTES

1. Valéry, who was a mathematician as well as a poet, wrote this in his notebook in 1935 (cf. Valéry, 1974: p. 811). Already in 1903 had he jotted down that "mathematics is the description of mental operations insofar as they can be exactly apperceived" (*ibid.*, p. 783).
2. If we are purists, we go to a dictionary to find the *correct use* – and we tend to forget that what we read there is still subject to *our* interpretation in terms of our own subjective experience.
3. For a full exposition of the radical constructivist theory of knowing see von Glasersfeld, 1978, 1981, 1985).
4. Starting from somewhat different considerations, Heinrich Bauersfeld has come to a similar conclusion and speaks of "subjective domains of experience" (1983).
5. E.g. Kilpatrick (1987), Kitchener, (1986).
6. The universally deplored decline of critical thinking ability is but one of the sad by-products of the exclusive emphasis on behavioral competence.

REFERENCES

Bauersfeld, H. "Subjektive Erfahrungsbereiche als Grundlage einer Interaktionstheorie des Mathematiklernens und & Lehrens". In H. Bauersfeld, *et. al.* (eds.), *Lernen und Lehren von Mathematik* (1–56). Köln: Aulis, 1983.

Kilpatrick, J. "What Constructivism Might Be In Mathematics Education". In J.C. Bergeron, N. Herscovics, & C. Kieran (eds.), *Psychology Of Mathematics Education, Vol. I* (Proceedings of PME-XI), 3–27. Montreal, Canada, 1987.

Kitchener, R.F. *Piaget's Theory Of Knowledge. Genetic Epistemology and Scientific Reason.* New Haven: Yale University Press, 1986.

Maturana, H.R. "Biology of Cognition." In H.R. Maturana and F.J. Varela (eds.), *Autopoiesis And Cognition.* Dordrecht: Reidel, 1980.

Piaget, J. *Six Psychological Studies.* New York: Vintage Books, 1967.

Rowell, J.A. "Piagetian Epistemology: Equilibration and the Teaching of Science", *Synthese* **80:** 141–162, 1989.

Valéry, P. *Cahiers, Vol. II.* Paris: Gallimard, 1974. von Glasersfeld, E. "Radical Constructivism and Piaget's Concept of Knowledge." In F.B. Murray (ed.), *The Impact Of Piagetian Theory* (109–122). Baltimore, MD: University Park Press, 1978.

— "Einführung in den radikalen Konstruktivismus." In P. Watzlawick (ed.), *Die erfundene Wirklichkeit* (16-38). Munich, Germany: Piper, 1981.

— *The Construction Of Knowledge: Contributions To Conceptual Semantics.* Salinas, CA: Intersystems Publications, 1987.

THERE IS NO MORE SAFETY IN NUMBERS: A NEW CONCEPTION OF MATHEMATICS TEACHING*

PHILIP H. STEEDMAN

University of Cincinnati

During the last half century profound changes in the way knowledge is conceptualized and evaluated have taken place. Work in the philosophy of science and in the psychological and sociological understandings of cognition itself have led to a basic reappraisal of scientific knowledge. Such a reappraisal could not but have repercussions in the field of education, even mathematics education. In this chapter I am reviewing general philosophical reasons that impinge on our conceptions of teaching and learning, as well as some changes of attitude that have already manifested themselves and others that are still in the making. Although I begin with considerations that are specific to the teaching of mathematics and end with an example that is more amply described in another chapter of this volume, my main focus is on the critical interaction between teacher and student and the school as the ambience where this transaction takes place.

We used to feel very confident about mathematics education. We knew what to teach, how to teach it, and why we should do so. Or we thought we did. Things are less certain today as the ground beneath the feet of all educators seems to move. Intellectual upheaval has not commonly been the lot of educators. Even if it had been, they could hardly have been prepared for what they are now experiencing.

The certainties of the past rested on three very grand sets of assumptions, commonly known as theories. The first addressed the "what to teach?" question. Mathematics was seen as a "discipline" and, in virtue of this, had an internal "logic" of its own, its own particular content, and boundaries which divided it from other disciplines. To be doing or teaching mathematics was not to be doing or teaching anything else. The content of such doing or teaching was determined not by individuals, nor in any ordinary sense by institutions, but rather by the discipline itself. Arguments of this kind have been mounted in various forms over the centuries; the most potent contemporary advocate of this view has been Paul Hirst who argues that the disciplines are distinguished from one another by their distinctive use of "categoreal" concepts. Given this, he argues that,

> It is these categoreal concepts that provide the form of experience in the different modes. Our understanding of the physical world, for instance, involves such categoreal concepts as those of *space, time, and cause*. Concepts such as those of *acid, electron, and velocity*, all

E. von Glasersfeld (ed.), Radical Constructivism in Mathematics Education, 1–11.

presuppose these categoreal notions. In the religious domain, the concept of *God or the transcended* is presumably categoreal whereas the concept of *prayer* operates at a lower level. In the moral area, the term *ought* labels the concept of categoreal status, as the term *intention* would seem to do in our understanding of persons. The distinctive type of objective test that is necessary to each domain is clearly linked with the meaning of these categoreal terms, though the specific forms the tests take may depend on the lower level concepts employed.[1]

The second theory addresses the "how to teach?" question. Here things seemed less clear but at least it was generally understood that students' learning was contingent on teaching (itself roughly defined as "acting with the intention to produce learning"). Students were seldom expected, like Descartes, to work things mathematical out for themselves but rather to have it "take" in their minds, or "pick it up" largely as a consequence of witnessing teachers' demonstrations and rehearsing examples. The situation, at least in outline, was quite simple. The teacher "knew the material", the students were "ignorant of the material", the object of the exercise of "teaching the material" was to have the students "know the material" by processes of learning. This was also, of course, the basis on which the teacher's authority was said to rest, an important point to which I will return later. The methods of teaching and learning reflected each other. Teachers demonstrated "content", usually of the theorem or proof sort. Students "worked" the same material. Learning was linked to horticulture as metaphors of tilling, planting, sowing and grafting were employed. Successful learning was detectable by doing: the student had "mastered" the material when he (or less commonly she) could "do" it.

Why one should teach mathematics had the grandest justification. There were various answers beginning with Plato's magnificent idea that the contemplation of mathematical knowledge not just puts the untrained mind in touch with what is timelessly real but also most perfectly develops the minds of future leaders.[2] Rejecting Plato, one might still have gained comfort from Newton's view[3] that nature has its own language and that is mathematics, that to unlock God's works is to know reality through numbers. Even rejecting Newton, one might have taken refuge in faculty psychology seeing the mind as some kind of metaphorical muscle to be grown through repeated exercise on very demanding material like mathematics. In this case the power to reason, developed mathematically (and, in the classical curriculum, by learning classical languages), could "transfer" to any other field, i.e., the person who could reason well in mathematics was seen as being capable of reasoning well in general.

Attacks on all three of these theories have been blistering this century. Thorndike attacked the classical curriculum with the empirical claim that "transfer of training" did not in fact occur and thus the foundation of faculty psychology was no more.[4] The onset of studies in the sociology of knowledge has shown it to be questionable whether disciplines do have lives of their own.[5] More recently still, Hirst's articulation of the forms of knowledge themselves has been attacked so thoroughly that it is doubtful whether it is now credible at all.[6] During these same decades a parallel development has taken place. It is the elaboration of a psychology of cognition in general which is an alternative to the notions of learning mentioned above. This work is most readily identified with the name of Jean Piaget, whose ideas shaped the background of most of the studies reported in this volume. Piaget's research program, of course, did not develop in a vacuum but rather proceeded at the same time as other, similarly fundamental work in the philosophy of science.

A revolution began with Karl Popper's *The Logic of Scientific Discovery*, which solidified the claim that it was no longer possible to defend any thesis which sought to make scientific knowledge certain. Popper argued that the problem of induction (the discovery of truths from collections of data) could not be solved in any traditional way. Scientific knowledge, therefore, could advance only by refuting rather than confirming hypotheses.[7] The effect was electric.

Norwood Russell Hanson[8] then focused the debate on the problematic relations between theory and observation. He pointed to the fact that no observation is possible without *some* theory, for without theory the observer would not know what to attend to. Clearly, Newton's "I make no hypotheses", and the neo-Newtonianism which has dominated Western thought ever since, was in serious trouble. Hanson's dictum that "there is more seeing than meets the eyeball" may as well have acted as the battle cry of a new school of post-Popperian philosophers of science. Popper had wanted to protect both the demarcation of science from other kinds of knowledge and the rationality of science by defining its method. This method of "falsificationism", according to which science proceeds by successively refuting hypotheses and theories and replacing them with new, more comprehensive ones, would indeed allow scientific knowledge to "advance", in an orderly way, a characteristic not shared by other sorts of knowledge. But this Popperian tranquility was short-lived.

Thomas Kuhn, in what must now be counted as one of the most influential books of the last half century,[9] proposed a new theory of scientific change. Science, Kuhn argued, does not proceed in an orderly way at all but rather by successive "revolutions". Building on Hanson's theory of dependence of observation, his own training in physics and his research on Copernicus, Kuhn saw two different sorts of sciences. "Normal science" was business as usual, the day to day practice of most working scientists who share a high degree of consensus about such issues as what is worth studying, how it should be investigated, what sort of instruments should be used, what an hypothesis is, and the like. But, in addition there is, Kuhn maintained, a "revolutionary science". Great changes occur in science not because of a gradual, systematic modification of world views but rather because new points of view come to restructure all aspects of "normal science". Copernicus and Galileo were revolutionary in this sense. Their work led to a new "paradigm" or way viewing the scientific enterprise.

The most shocking element of Kuhn's work, however, was the notion that paradigms are often incommensurable, i.e., they cannot be compared because each involves concepts that are incompatible with the other. The implication was shattering for traditional views of science. If there is no neutral, third position from which to compare two competing paradigms there is no traditionally "rational" way of selecting one over another. Not only were Copernicus and Galileo not "better" than Ptolemy from the point of view of some time-less, value-free, God's eye position, but there is *no such* position. Copernicus and Galileo "won" over Ptolemy because of factors outside science; Ptolemy had become unfashionable, Galileo was seen as "the wave of the future". Thus Popper's attempt to demarcate science from the rest of human life was defeated.[10]

These developments reached an at least provisional culmination in the work of Paul Feyerabend, a student of Popper and colleague of Kuhn at the University of California at Berkeley. Feyerabend denied that there was anything special about science at all. It has no privileged "method", no special "objectivity", no "progress", no grounds at all to claim to be better than other forms of knowledge or ways of life.[11] Attempts to assert such claims are always rationalization *after the fact*, rationalization that serves to mask what is really going on. Feyerabend is often understood as being "anti-science". This is false, and importantly so. He merely argues against the institution of scientific knowledge as unquestionable dogma.

> ... assume that the pursuit of a theory has led to success and that the theory has explained in a satisfactory manner circumstances that had been unintelligible for quite some time. This gives empirical support to an idea which to start with seemed to possess only this advantage: It was interesting and intriguing. The concentration upon the theory will now be reinforced, the attitude towards alternatives will become less tolerant.
> ...At the same time it is evident, ... that this appearance of success *cannot be regarded as a sign of truth and correspondence with nature*. On the contrary, the suspicion arises that the absence of major difficulties is a result of the decrease of empirical content brought about by the elimination of alternatives, and of facts that can be discovered with the help of these alternatives only. In other words, *the suspicion arises that this alleged success is due to the fact that in the process of application to new domains the theory has been turned into a metaphysical system*. Such a system will of course be very 'successful' not, however, because it agrees so well with the facts, but because no facts have been specified that would constitute a test and because some such facts have even been removed. Its 'success' *is entirely man-made*. It was decided to stick to some ideas and the result was, quite naturally, the survival of these ideas.[12]

Feyerabend's respect for science, and indeed all knowledge, has led him to want science understood for what it is, rather than being misunderstood as a secularized religion.

This change of attitude has begun to affect not only the scientific disciplines but also the theory and practice of education and, in the last decade, specifically mathematics education. To appreciate the present changes in that particular field, however, we must look at what is happening to the general notions of knowledge and the ways and means of fostering its acquisition. It is an index of how powerful recent theories and ideologies of science in particular have been in influencing all varieties of knowledge, that the *general* debate about contemporary epistemological questions in all fields should have been so deeply influenced by the writers mentioned above. And, sadly, it is a reflection of the abject poverty of much that is contemporary educational theory and philosophy, that these recent developments are so seldom mentioned, let alone debated.

Educational thought has functioned for decades as little more than a watered down branch of psychology, and indeed that most mindless variety of psychology which has seen a limp behaviorism as an appropriate means for understanding the development of cognition. The nature of knowledge and its development by human persons and institutions has commanded relatively little serious attention by educational thinkers. Hirst's discipline model of the curriculum has held sway despite decimating attacks.[13] Piaget's majestic work has been "psychologized" by ignoring its genetic epistemology and focusing on "stages of development" with an attendant elaboration of ideas of "readiness". By these means a bizarre type of Platonism has been perpetuated. One can only hope that the appearance of this, and similar works, signal an end to this situation.

It is now time to face what we do know, and it is this. Educational situations are ethical in their essence. The term "individuals", with its suggestion of neutral objectivity, must therefore give way to "persons".[14] This used to be a commonplace in earlier times, one which hardly needed to be pointed out and explicated but was taken for granted. To see that this is not the case today is to notice how far our culture has traveled away from meaningful moral discourse. "Individuals" carries a psychological connotation, not a moral one. Psychologists during the last hundred years have succeeded to an alarming extent in redefining the nature of human life, away from the earlier conceptions of us as moral, aesthetic, and political, to conform to the mechanistic world view which had been propagated about the universe since Galileo and Newton. A technology of social and person control has developed around these ideas (see note 16). Freud, Skinner, and others have sought to convince or persuade us that we are no more than the passive products of genetic or environmental and historical circumstances. The implication is that we can and should be molded by "therapeutic intervention" or "social planning". This sort of thinking has been fostered by a devastating parallel which it then itself encouraged: that we can reasonably consider our lives in a morally neutral context. Thus questions of "psychological health" are seen as *technical* and have to a great extent replaced moral questions. It is not that the latter could not be answered – it is that they hardly make sense in this intellectual climate. Thus education, which is *clearly* to mean improvement, moral improvement, through the development of cognitive autonomy, was emasculated to mean something like "any learning programmed by the school" or even simply "any learning" (see Peters, note 14). This moral neutrality has the effect of trivializing our experience of ourselves. We *can*, of course, consider ourselves as machines if we choose to. There are circumstances where this makes good sense: for example when we are on the surgeon's operating table. But such thinking can hardly be appropriate in all cases: we certainly do not want *the surgeon* to be a machine during our operation, we want moral consideration and we are right to want it.

A very similar situation obtains with teaching and learning. Persons learn not by being given knowledge but rather by constructing knowledge. So the "banking" model of teaching as making "deposits" in the mental accounts of students must be rejected. Rather, the teaching and learning situation must be understood in the following way. First, since we have no privileged access to reality, there is no template of that reality which *could* be "deposited", even if we thought such teaching desirable and even if students had an "open account" awaiting such deposits. Second, learning, even learning physical habits such as cleaning one's teeth and tying one's laces, always involves a subtly complex negotiation between teacher and student in which both are changed. For the student's part we must recognize that the learning involves building up knowledge with the means and the material accessible to the student, rather than copying reality. This is another place where the arguments from the contemporary philosophy of science are so persuasive (see notes 7–10). "Reality" is not sitting "out there" awaiting appreciation through neutral "observation"; observation is always a theoretical activity carried out by an observer. Nature (itself a theoretical construct) is not waiting to whisper its secrets in our scientifically cleansed ears. Theories are not mental representations of reality (despite the seduction of that word *representations*). To observe is to interpret. This does not mean that there is no way we can evaluate such activities. Abandoning the "theory as template of reality" account of things releases us to acknowledge that we could judge the quality of such activities on *whatever relevant grounds we choose*. Strong cases can be made for candidates such as purposes (what is the knowledge for?) and moral/social considerations (will this knowledge help us to lead more humane lives?), but the list is endless. The implication is that there is no single "right" or "correct" way of understanding, and it is time we embraced rather than continued to resist it.

The pedagogical implication is no less far reaching: not everyone learns in the same way. Of course, we all know this to be true but it is a truth which was blurred by focusing on psychological measurement (or, more often mismeasurement) of "individual differences" in order then to ignore, for the most part, what such measurements might reveal. In other words, there has been a peculiar mismatch between our notions of the nature of the learner, the activity of teaching, and the nature of knowledge.

This situation has been perpetuated by the third truth to be faced: situations specifically designed for learning (i.e., "schooling") are the sites of powerful social forces which shape *all* aspects of what goes on within them. Who and what teachers are is shaped by these forces; how the learner is understood and treated is shaped by these forces; and what is taken to be "knowledge" or "worth knowing" is constituted by these forces.[15] For those who value freedom and autonomy we can hardly conclude that these forces are universally benign.[16]

It is not just a series of accidents of misunderstanding which have led to the situation we find ourselves in today. Rather, we find schooling serving certain social interests, its self-understanding and practices operating to modify, create, and legitimize some special arrangements and not others. One of the most significant aspects of how these processes work, both within schools and between schools and other social institutions, is that it is often claimed that such processes *do not exist*. Alternatively, it is claimed that they are an unfortunate and insignificant byproduct of the institutionalization of schooling. Schooling, at its heart a political enterprise, is thus *apparently* depoliticized. But teaching by its very nature is a political activity; insofar as it is successful it fosters the construction of a specific reality and could hardly, therefore, be a more pervasively political activity.

We should now return to the three sets of assumptions with which we began. Mathematics is not a discreet and separate enterprise unrelated to other varieties of knowledge and action. It is a social creation which changes with time and circumstances. All sorts of criteria influence these changes, not the least of which are economic, political and aesthetics. There is no timeless mathematics standing outside history which could be taught. If this seems counter-intuitive then we must face the fact that our intuitions have been well trained.

The point is proven by reference to cross cultural studies. Needham[17] has shown that the Chinese, despite developing a mathematics and physics of great sophistication, never invented the calculus or atomic theory. Nature (and numbers) is not "out there" exerting some kind of mental magnetism which attracts us to "truth". And within our own culture examples in the history of Western science abound. Galileo's telescopic research was influenced, and deeply, not just by factors internal to science but also by the needs of Venetian merchants wanting to manipulate market prices, who took advantage of the new "fairground toy" (the telescope) to see the cargos on the decks of ships before they docked. Kepler's magisterial work would have been briefer had he not been wedded to the notion of spherical celestial motion for religious and aesthetic reasons. Physics, as we understand it, developed in part because the British Admiralty needed an accurate clock, not to promote science but rather to ensure precise navigation; they were in the business of trying to protect shipping in a growing colonial empire.

Secondly, "depositing" is no metaphor of teaching. The teacher who cannot tolerate different ways of knowing is doomed to failure. Most worry about this point seems to spring from a concern for the teacher's authority; if there is not one "correct" way to teach, what authority does the teacher have? But such a concern rests on a misunderstanding of the question which confuses being "an authority" with "being authoritarian." Abandoning authoritarianism does not entail abandoning being an authority. Much of traditional mathematics teaching has relied on the former, and this has led to many of the problems in the field today. The mathematics teacher's authority rests on expertise and maturity, not on the arbitrary exercise of power. So what might an alternative understanding, at the most general level, be?

It is refreshing for a philosopher to be able to report in answer to this question that something a good deal more concrete than speculation may now be offered. This very volume comprises several examples of the sort of alternative practice I am advocating to replace the dismal procedures based on the obsolete assumptions outlined above. One of these examples is reported in Paul Cobb, Terry Wood, and Erna Yackel's paper "A Constructivist Approach to Second Grade Mathematics". I want to draw your attention to certain features of this work, some more obvious than others, and then draw some general conclusions. But I should point out at once that I consider it no more important or more significant than other contributions to this volume. I am using Cobb *et al.* as a point of reference because I had the opportunity to read this chapter and to discuss the work with Paul Cobb before I wrote my own piece.

The first thing to note is that this is conceptually well grounded work. Cobb *et al.* have addressed the central core of literature in not only contemporary cognitive psychology but also contemporary philosophy of science.[18] In other words it is work which, despite the modest tone of its reportage, has broken out of the normal framework of discourse in mathematics education and is in fact *radical* in the literal sense of that word, it returns to fundamentals. It is no wonder, then, that its results are such dynamite! Secondly, it is real *schooling* research that is reported here. Cobb and his co-workers have not been satisfied with what normally passes for "educational research", i.e., the "laboratory trials" or the "lab school". Rather they have sought to situate their work in what is, at once, the most exciting, challenging and relevant of all settings, a real school with all that entails: real pupils, real grades, real teachers, real parents, real administrators, real school board, etc. It also means that the results described in this report have real credibility.

It is clear that questions of authority are central in this work. There are two major areas. First, there are questions related to what we might describe as the "authority of the material". Second, there are questions related directly to the teacher's authority. Clearly these interact in powerful and subtle ways. In the first case Cobb, his collaborators, and other similarly oriented researchers have abandoned many of the safe notions on which much mathematics education has rested. While it is still individual persons who are learners, they are understood as participants acting and learning within the social constraints of a classroom. Both what is learned and how it is learned are contingent to some extent on a redrawing of the boundaries between the individual learner, the teacher, and the whole group (including the teacher). This, of course, has a direct impact on the authority of the teacher.

That this is inevitable only becomes the more obvious when we note how some of the other "safe assumptions" have also been abandoned. In this classroom there is no one way to "best" solve a problem. The teacher really does not "know best" all the time. But further, there are no "problems" in the traditional sense either. The teacher does not pluck "problems" from the "logical structure of mathematics itself" and try to "insert" them into pupils "minds". Instead, the problems usually arise from within the experience of the pupils and the particular social setting generated in the classroom. The classroom is a constrained situation; people are there for reasons; work has to be done; the teacher is "in charge", is responsible and knows more. But none of this entails the traditional mathematics education practice or its justifications. The irony here rests in the fact that the traditional authoritarianism of mathematics teaching was phoney whereas the authority exercised in Cobb's *et al.* classroom is real just because it depends on a proper recognition of the individual's construction of knowledge within a social setting and the practical requirements of maintaining such a setting.

This work (and indeed this whole volume) raises a number of other important issues only the most general of which I can address here. I want to call it the issue of credibility. The authors seem to assume that the reason schools attempt to teach mathematics is that they want all children and adolescents to learn mathematics, that a mathematically competent general population is a long-range social goal of contemporary schooling practice. But there is now a daunting and unignorably potent literature which suggests that public schooling does not exist, and was never established, to promote the fullest cognitive autonomy for the whole population. In the light of recent historical and sociological analysis this seems simply another false assumption.[19] Just as schools do have an educational mission they also have what has been described as a "cooling out" one too.[20] They are in the business of selecting, classifying, certifying and manipulating the expectations of their pupils *in both directions*.[21] Schools are, to put the matter plainly, sites of social conflict[22] and these conflicts shape the contours of the educational and miseducational activities in which they engage. All this is as true of the preschool and elementary school as it is of the rest of the system.[23] Of course to point this out is not to offer a critique of the reportage by Cobb *et al.* It is to note that work of this sort will finally need to be situated in a more general theory, first of schooling (which links curriculum, teaching and learning with the political discourse of which they are in fact part) and, secondly, with a theory of education (in which the ethical structure of schooling is comprehended).

The work I have cited implies such a direction. Its authors, as do others who contributed to this volume, force us to think through an alternative epistemology (which is a constructivist one) and pedagogy. This shift of perspective suggests that we should teach mathematics for many reasons. Beyond the obvious ones (getting a job, etc.) mathematics is essential to the contemporary sciences. But there are at least two further reasons, and they are subtly connected. One is that mathematics is part of that great treasure trove of human understanding which this culture has constructed in its history and by which we can know ourselves. Like the works of the finest poets, philosophers, artists, and scientists, the creations of mathematicians show us the human potential (at least in the West) operating at full strength. No less is the educational right of every student. The other is that mathematics is fun. Or at least it can be. Why it is often not, should exercise the wits of all those who seek to understand, not just mathematics teaching but the phenomenon of schooling itself.

NOTES

* Paul Feyerabend and Ernst von Glasersfeld read drafts of this paper. I am grateful to them both for their suggestions and to the latter for his gracious editing.

1. See the now classic "Liberal Education and the Nature of Knowledge" and Hirst's other related papers in P.H. Hirst, *Knowledge and the Curriculum*. London: Routledge and Kegan Paul, 1974.
2. See Plato's *Republic*.
3. See F.E. Manuel, *A Portrait of Isaac Newton*. Washington, D.C.: New Republic Books, 1979.
4. See the treatment of Thorndike in C.J. Karier, *Scientists of the Mind*, Urbana, IL: University of Illinois Press, 1986.
5. See my "Curriculum and Knowledge Selection" in L.E. Beyer and M.W. Apple (eds.), *The Curriculum: Problems, Politics and Possibilities*. Albany: SUNY Press, 1988.
6. *Ibid.*
7. See K.R. Popper, *The Logic of Scientific Discovery*. London: Hutchinson, 1959. *Conjectures and Refutations: The Growth of Scientific Knowledge*. London: Routledge and Kegan Paul, 1963. *Objective Knowledge: An Evolutionary Approach*. Oxford, U.K.: Oxford University Press, 1972.
8. N.R. Hanson, *Patterns of Discovery*. Cambridge, U.K.: Cambridge University Press, 1972.
9. T.S. Kuhn, *The Structure of Scientific Revolutions* (Second Edition). Chicago: University of Chicago Press, 1970.
10. This argument is elaborated in Imre Lakatos' classic "The Methodology of Scientific Research Programmes" in his *Collected Papers, Vol. I*. Cambridge, U.K.: Cambridge University Press, 1978.
11. Feyerabend's writings are now voluminous but see P.K. Feyerabend, *Against Method: Outline of an Anarchistic Theory of Knowledge*. London: New Left Books, 1975. *Science in a Free Society*. London: New Left Books, 1978. *Collected Papers* (two volumes). Cambridge, U.K.: Cambridge University Press, 1981. *Farewell to Reason*. London: Verso, 1987.
12. P.K. Feyerabend, "How To Be A Good Empiricist – A Plea For Tolerance In Matters Epistemological". In P.H. Nidditch (ed.), *The Philosophy of Science*. Oxford, U.K.: Oxford University Press, 1968.
13. It is ironic that Hirst's *magnum opus (op. cit.)* was published during the period of most furious debate over Kuhn and Feyerabend's work, yet they are not so much as mentioned by Hirst. See my "Should Debbie do Shale?", *Educational Studies*, Vol. 13, No. 2, Summer, 1982.
14. See R.S. Peters, *Ethics and Education*. London: George Allen & Unwin, 1966.
15. For excellent introductions to this literature which is now vast, see J.A. Scimecca, *Education and Society*. New York: Holt, Rinehart and Winston, 1980, and J. Spring, *American Education: An Introduction To Social and Political Aspects* (Third Edition). NY: Longman, 1985.

16. The relations between social forces, the constitutions of persons and institutions, and the nature of knowledge is discussed in detail in my *The Educational Panaticon: Schooling After Foucault*, (in progress).
17. Needham and W. Long, *Science and Civilization in China, Vol. 4, Physics and Physical Technology Part I, Physics*, Cambridge, U.K.: Cambridge University Press, 1962.
18. Paul Cobb, private communications 1986 and 1988. (The sort of literature I have in mind is represented in Notes, 7, 8, 9, 10 and 11 above).
19. See C.J. Karier, *The Individual, Society and Education: A History of American Educational Ideals* (Second Edition). Urbana, IL: University of Illinois Press, 1986.
20. See B.R. Clark, "The 'Cooling-Out' Function in Higher Education", *American Journal of Sociology*, May, 1960; 569-76.
21. See P.C. Violas, *The Training of the Urban Working Class*. Chicago, IL: Rand McManny, 1978.
22. See J. Spring, *Conflict of Interests: The Politics of American Education*. NY: Longman, 1988.
23. See M.W. Apple, *Ideology and Curriculum*. Boston, MA: Routledge and Kegan Paul, 1979 (especially Chapter 3).

MATHEMATICAL DISCUSSIONS [1]

JOHN RICHARDS

BBN Laboratories, Cambridge, MA

> *As civilized human beings, we are the inheritors, neither of an inquiry about ourselves and the world, nor of an accumulating body of information, but of a conversation, begun in the primeval forest and extended and made more articulate in the course of centuries. It is a conversation which goes on in public and within each of ourselves.... Education, properly speaking, is an initiation into the skill and partnership of this conversation in which we learn to recognize the voices, to distinguish the proper occasions of utterances, and in which we acquire the intellectual and moral habits appropriate to conversation. And it is this conversation which, in the end, gives place and character to every human activity and utterance.*
>
> Michael Oakeshott [1962], p. 199

I. THE DIALOG – AN ATTEMPT AT INQUIRY TEACHING

Two tenth grade students in a remedial tutoring session are working on an algebra problem presented in the *Algebra Workbench* (see figure 1).

$$6n + n = 21$$

Figure 1. The Problem.

The *Workbench* is a piece of computer software designed to simplify the acquisition of strategic problem solving skills by separating them from the operations of algebraic manipulation (cf. Richards and Feurzeig [1988]). The hope is to let the software do the computation so the teacher and student can focus on planning and strategy. The following dialog occurred in the first session after introducing the students to the software.

Teacher: *So, what do you want to do with the equation?* (Pause ... maybe fifteen seconds, staring at the screen.)

J: *Solve it!*

E. von Glasersfeld (ed.), Radical Constructivism in Mathematics Education, 13–51.

Teacher: (Pause ... maybe five seconds.) *Um, uh, that's good. How would you want to solve it? What can you do to the equation?* (Dead silence – after about fifteen seconds.) *What are you thinking about?* (Dead silence – another ten seconds.) *What steps do you ...*

G: *Reciprocal!*

Teacher: *What?*

G: *Reciprocal!*

Teacher: *What do you mean?* (Pause ... no answer.) *What can you do with the reciprocal? How can you use it?*

G: *You multiply by 1 over 6.*

Teacher: *Oh, you mean you divide by 6?*

G: *You do?*

As the teacher, I was startled by J's "Solve it!" It occurred to me, at first, that this was sarcastic, but I quickly rejected the idea. The students were very uncomfortable, but they were eager to please. J was serious, but she had no way to understand what I wanted. In retrospect, this was the first small step toward communication, but at the time all I thought was how far we needed to go. The students had been given some mathematical words, and G uttered one of them, "reciprocal," hoping that this would satisfy me. She only uttered the word in isolation, not embedded in a sentence. Clearly it is only vaguely related to any deeper understanding of the concepts involved. Note also, that, although this is not a wrong answer, working with the reciprocal complicates the problem and doesn't lead to an optimal solution.

This episode is from a project that attempts to shift emphasis away from teaching mathematics as a body of facts and procedures. Students are introduced to the activity of mathematics as informed exploration and reflective inquiry. I was trying to use the computational environment to teach the skills of mathematical inquiry. I had anticipated a discussion on planning and developing strategies for solving equations. The girls were focusing on entirely different issues. They were trying very hard to connect to whatever it was that was happening, and it was very different from any school experience they knew.

The episode is marked by the inability of the students and teacher to communicate. It is not as if these particular students were non-verbal. In fact, they were quick to talk in almost every other circumstance, and when they were asked to write about their feelings regarding mathematics, they easily filled one and a half pages. Rather, the teacher and students were operating with different assumptions about their respective roles, different expectations about the purpose of their interactions, and different understandings about the nature and function of mathematics.

II. OVERVIEW

Mathematics is a socially constructed human activity. It has a history, a tradition, and a culture. Yet each individual constructs his or her own mathematics. Typically, constructivist accounts focus on the latter activity by, for example, modeling the individuals' construction, or analyzing the nature of the construction itself.[2] While I do not wish to minimize the individual's construction, in this paper I look at a social aspect – learning in the schools. I provide a framework to examine the process whereby a teacher and student interact.

The teacher and students in the above dialog were not communicating, but they were exchanging words. They were not communicating, because they assumed roles in very different mathematical communities, and might as well have been speaking foreign languages to each other. Mathematics is the subject matter, or content, for at least four distinct domains of discourse. There are different cultures associated with each of these different domains, with different expectations regarding the assumptions, goals, and underlying methodologies. Moreover, there is a different mathematics associated with each of these linguistic domains. In many ways the domains are analogous to different methodological research programs in science (cf. Lakatos [1970]).[3]

1. **Research Math** – the spoken mathematics of the professional mathematician and scientist. Professional mathematicians share much of their discourse structure with other research communities in that the language is technical and makes many assumptions regarding the underlying content area. For the purposes of this paper, I include scientists and engineers within this community, although it is clear that there are important distinctions between these groups. The mathematical research community discourse is structured according to a "logic of discovery" (cf. Popper [1959]) stressing the actions of making conjectures and refutations. The characteristic that distinguishes mathematicians from other research communities is their subtle reliance on notions regarding the nature of proof.
2. **Inquiry Math** – mathematics as it is used by mathematically literate adults. This approximates, but is still distinct from the language of the "mathematical research community." The language of mathematical literacy includes participating in a mathematical discussion, and acting mathematically – asking mathematical questions; solving mathematical problems that are new to you; proposing conjectures; listening to mathematical arguments; and, reading and challenging popular articles containing mathematical content.
3. **Journal Math** – the language of mathematical publications and papers. The emphasis is on formal communication, at a distance or across time, where there is no opportunity to clarify ambiguities. This language is very different from the spoken language of the research community, that is, it is very different from a logic of discovery. Papers and publications are based on a

"reconstructed logic" that makes mathematical discoveries more palatable for public consumption (cf. Richards [1988]).

4. **School Math** – the discourse of the standard classroom in which mathematics is taught. This discourse does not differ much from "school science" or "school English." Viewed as discourse structures, classroom lessons mostly consist of "initiation-reply-evaluation" sequences (cf. Mehan [1979]). What is learned is useful for solving habitual, unreflective, arithmetic problems. In many ways, this discourse does not produce mathematics, or mathematical discussions, but rather a type of "number talk" that is driven by computation.

There are important analogies between *research math*, the language of the mathematical research community, and *inquiry math*, the language of a mathematically literate adult. Both domains are used in a "logic of discovery." In contrast, *school math* is patterned after the mathematics of journal articles. Journal articles, with their dependence on "reconstructed logic," are designed to transfer information to a community that has already accepted many presuppositions and constructed a great deal of mathematics. This is useless for students who are faced with the need to construct these presuppositions and to construct mathematics for themselves.

The students in the first dialog were speaking *school math*. In contrast, the teacher was speaking *inquiry math* – the language of mathematically literate adults. The teacher in the interaction was operating from a different model of mathematics than is found in most classrooms. The students had little or no experience with mathematically literate adults, and did not know how to participate in a mathematical discussion or to act mathematically. This, and similar episodes, suggests that in their previous *school math* mathematical experience these students had been treated as passive recipients of information. They were not prepared to inquire about mathematical issues, or assume responsibility for their own learning. These are the most obvious negative implications of *school math*, that is, of years of having ignored the students' need to construct mathematics for themselves.

There is a second, deeper, consequence of restricting school mathematics to the *school math* conversation. As Oakeshott argues in the opening quotation, education initiates students into the conversation of humanity. This conversation goes on in public, and is learned by participating in a public conversation. Ultimately, we must be able to carry on this conversation by ourselves. By continuing the conversation within ourselves we begin to act mathematically. Our own conversation then serves as an aid in posing and solving problems. Our ability to continue the conversation gives us the power to engage mathematical issues. We first learn to continue the conversation by ourselves by participating in conversation with others.

Obviously, *school math* is the wrong math to be teaching students. The current literature is replete with proposals to revise the context for learning and doing mathematics, such as pair problem solving (Lochhead, [1985]), inquiry learning (Lampert [1986]), modeling mathematical behavior (Schoenfeld [1983a]), reciprocal teaching (Palincsar and Brown [1984]), and cognitive apprenticeship (Brown and Collins [in press]). These proposals are oriented towards problem solving, and trace their intellectual lineage through Polya and Dewey. In this paper I argue that what these diverse teaching methodologies have in common is their attempt to change the language of the classroom from *school math* to *inquiry math*. The need for students to construct mathematics for themselves can be taught best in an atmosphere where students can talk *inquiry math*, that is, a classroom designed to teach the language of mathematical literacy.

This paper grounds an inquiry approach to teaching and learning in a constructivist epistemology by focusing on the area of interaction between teacher and student. This interaction occurs in discourse. For mathematical inquiry to take place students and teachers must learn to carry on a mathematical discussion – they must learn to speak *inquiry math*. I am not suggesting that every member of the pre-college school population become a member of the mathematical research community. Rather, schools must provide a means for students to become mathematically literate adults. They must give students the language and the tools to continue the conversation for themselves.

III. A CONSTRUCTIVIST ACCOUNT OF COMMUNICATION

How does an individual learn to communicate with other mathematically literate adults, and insofar as there is a community of mathematically literate adults, how does an individual become a member? This is both an epistemological question and an ethnographic, descriptive, question. In order to address these distinct aspects, this section surveys two extensive and very distinct bodies of research. I have tried to weave these together, but mostly this is intended as a pointer to some seminal pieces in the literature.

First, from a constructivist perspective, how is it possible to communicate with any other individual, and to become a member of any community? We need to establish an epistemological basis for interaction and communication. If all knowledge is constructed by an individual, how do we share knowledge, and how do we create knowledge in common?[4] Most important, for educators, how can one individual, a teacher, act to change another individual, a student? The writings of Humberto Maturana provide an account of how I, as an observer, can productively describe the process of human interaction.

Second, from an ethnographic perspective, we need to establish a linguistic basis for analyzing the dynamics of the classroom discourse. Ethnographic and sociolinguistic studies of classroom behavior focus on discourse as the fundamental linguistic unit of analysis. These studies have carefully defined terms in order to provide a means for researchers to talk about discourse. In this section I briefly describe some of this language to use for our own purposes in the rest of this chapter. Ordinary language is fraught with realist undertones, and as a constructivist one needs to be very careful about choosing, and clarifying a language for descriptions.

3.1. Consensual Domains

Discourse can occur because people, as organisms, have established, in Maturana's terms, a "consensual domain" (cf. Maturana [1978] pp. 47 *ff.* and Winograd and Flores [1986] pp. 48 *ff.*). A consensual domain is a fundamental biological relationship, and is established when two or more organisms interact recursively. Each organism acts in response to the other. This can be seen on a prairie, for example, when two tree seedlings grow together. From a distance, to an observer, they may appear as a single tree, each occupying the space left by the other. In the interactions between them each organism undergoes a process of structural coupling due to the perturbations generated by the other (Winograd and Flores [1986], p. 48). The conduct of each organism is interlocked with the other. The domain of interlocked conducts is a consensual domain.

Since several organisms can participate in a consensual domain, some may pass in or out of it. The consensual domain may continue even if one of the members no longer participates. Thus, a consensual domain may take longer than a lifetime to develop, and may evolve through many lifetimes. Many of the participants who worked to form a consensual domain may be dead (and thus no longer contributing to it). It is in this sense that Oakeshott (in the opening quotation) can speak of a conversation of humanity, that has continued over time.

For Maturana, behavior in a consensual domain is linguistic behavior. Language, as a consensual domain, is a patterning of "mutual orienting behavior." The organisms have reached a consensus in their behavior. They have not necessarily consciously (or reflectively) *agreed*, but rather, from the perspective of an observer, they are acting in accord. They are acting *as if* they have come to an agreement regarding their underlying assumptions. Focusing on establishing a consensual domain underscores that it is the process that we are interested in, not the product (whatever that might be). It further emphasizes that the formation of a consensual domain requires great effort.

Halliday describes the process of acculturation in a similar style:

> This does not happen by instruction ... nobody teaches him [the child] the principles on which social groups are organized, or their systems of belief, nor would he understand it if they tried. It happens indirectly through the accumulated experience of numerous small events, insignificant in themselves, in which he contracts and develops personal relationships of all kinds. All this takes place through the medium of language. (Halliday [1978], p. 9)

Becoming a member of a community is a gradual process of mutually orienting linguistic behavior.

School math is a product of years of reflexive activity between students and teachers. From an observer's perspective each of their behaviors is coupled to the other. The initiation-reply-evaluation pattern mentioned earlier is a highly evolved consensual domain involving lots of other behaviors (e.g., taking turns, raising hands). This reflexive activity constitutes or shapes a consensual domain.

The mere saying of words does not make a conversation, nor does it make for communication. The words must occur in context, and context is established in a consensual domain. Being in a consensual domain is a necessary, but not a sufficient, condition for successful communication. There is a consensual domain established in schools, but it is not one in which communication occurs.

3.2. Conversations in Which Communication Occurs

For Maturana, communication takes place after the establishment of a structural coupling. From a discourse dynamics perspective, structural coupling is a reflexive activity:

> When people are ... in face-to-face interaction, what looks on the surface to be a series of discrete, successive "turns" is actually a process of continuous, simultaneously reflexive behaving and monitoring by the two players. Through such reflexivity, the conversation can be said to be *jointly produced* by its participants. (Erickson [1980], p. 8)

Through reflexivity, participants establish that they are in the same ball game. Only then can we begin to speak about communication.

If reflexive activity leads to a consensual domain, it is, in the strict sense, a conversation, "a turning around together" in such a manner that all participants undergo nontrivial structural changes until communication takes place.[5] Pre-communicating interactions, which take place during a conversation, are creative interactions. These mutually orienting activities lead to novel behavior (Maturana [1978] pp. 54–5).

For communication to occur, both participants must have the potential for change. In the weeks following the dialog at the beginning of this paper, the students learned how to communicate with me. In addition, I learned to communicate with them. This *mutual* adjustment created a new context in which communication was possible. As we established a consensual domain we created a foundation, a mutual understanding, that allowed us to communicate. The participants in a consensual domain learn what they can take for granted, what is obvious. Only after this foundation is established does communication begin. We only say something when it is not obvious. A consensual domain permits us to assume that much is obvious.

Misunderstanding can, and often does, occur within a consensual domain. In communicating an idea there is a give and take, there are questions, answers, doubts, challenges and compromises. These occur between communicating members of a consensual domain. This is what is meant by negotiating meanings.

From the ethnographic perspective this process is understood as occurring in a text. A text is a linguistic interaction, either spoken or written, considered as a unit.

> [S]peaking is not simply a matter of individuals saying what they want, when they want to say it. Conversational cooperation is managed .... This ... process is affected by extralinguistic constraints and by semantic ties which cut across utterances. (Gumperz [1982], p. 11)

A text can be analyzed in terms that take into account the dynamics of interaction, such as speech acts, turns, and topics.

A text has structure over and above the individual utterances of sentences that are contained within it. In ethnographic terms, text is a semantic unit that incorporates a social milieu. By taking text as our basic semantic unit, we are changing the level of analysis away from the word or sentence, or even the speech act. Words have different meanings in different texts. Wittgenstein's claim is fundamental: a word's meaning is in its use. This should not be taken narrowly as requiring that we consider words juxtaposed into sentences as having meaning. For words or sentences to have meaning they must occur in texts.

3.3. Subdomains

An adult member of society plays many roles, i.e., constructs consensual domains with many individuals. These constructions all occur through the medium of language. A speaker of English does not automatically communicate with every other speaker of English by the mere process of knowing the language, or of having established consensual domains with other speakers of English. Consensual domains are not transitive, in the sense that participating in a consensual domain with someone does not necessarily bring you into other consensual domains in which that person participates. Moreover, even when you participate in a domain of discourse, there may be distinct subdomains of discourse that have adapted their own specialized vocabulary, syntax, and pragmatics and thereby created a distinct consensual domain. Participation in an overarching domain does not guarantee participation in subdomains.

The construction of distinct consensual domains corresponds to the observation that the language we speak or write varies according to the type of situation. We experience language in context, in a scenario, and not in isolation. In ethnographic terms, we say that there are distinct registers. A register is comprised of the type of event in which the text is functioning, the purposive activity of the speakers, the content and function of the text, genre, rhetorical mode, types of role interactions, relevant social relations, and so on (cf. Halliday and Hasan [1976], p. 22 *ff.*). In this sense, the distinct mathematical domains of discourse, outlined above, are four different registers.

The participants in the first dialog have not established a consensual domain. They are only beginning a conversation, and are clearly not communicating.

There is a clash of registers. Both the teacher and the students have made certain assumptions about the nature of the discourse. I was not in the standard teacher mode. And as the students realized this, they tried to figure out who I was, i.e., what role was I playing. This became clear when, in our second session, G asked me, "Can I have a lollipop?" She would never have asked that of a regular teacher. (I said "Sure." And her response, "It makes me think better," was very interesting in the context of what I was trying to accomplish.)

3.4. Rhythm

Changing the focus of our ethnographic investigations from sentences and meanings, to texts and discourse is a result of changing the pedagogical focus from a concern with products to a concern with process. But evaluating a process is a very different kind of activity from evaluating products. Social interactions are complex, and this complexity is reflected in the difficulty of establishing clear criteria by which we can judge their success or failure. A clash of registers, as in the dialog, is usually pretty obvious, and marked, not coincidentally, by the participants feeling uncomfortable, awkward, or ill-at-ease. Even within the same register, there are subtle failures of communication, which are based on narrower presumptions made by the participants regarding shared beliefs and assumptions about the context (time, space, history). Teachers and researchers are faced with the additional difficulty of trying to apply these criteria at the same time they are trying to communicate.

As a participant, one is often embarrassed when watching replays of taped sessions. After seeing a scene several times on videotape, you realize that your understanding of the events at the time of the session was significantly different from what you see later in the replays. Reports on the time-intensive and controversial nature of these sessions are common from projects that rely on videotaped data.[6] It is no surprise that there are misunderstandings in real human interaction.

When you examine a text from a social perspective a message is evaluated in terms of its "fit" or "lack of fit," rather than being "right" or "wrong." Success in these terms is measured by the "appropriateness" of the message (cf. Erickson [1980], p. 5). In occupying diverse consensual domains, we play many different roles simultaneously. The appropriateness of a message is determined by the respective current roles of the speaker and listener.

> While the speaker or speakers are doing speaking, the listener or listeners are doing listening. Listeners' ways of doing listening apparently provide speakers with information about how the spoken message is getting across, and that information is apparently used by speakers in shaping the recipient design features of their speech as they are talking. (Erickson, [1980], p. 8).

Listening or not listening, paying attention and nodding agreement, or not paying attention and not waiting for answers to questions, affects the rhythm of the conversation. The speaker and listener become attuned to each other, and pick up on each other's cues. Timing is an essential ingredient in generating communication. Communication is socially and rhythmically organized (Erickson, p. 37), and speaking and listening behavior occurs in synchrony (p. 9). There are other sorts of obvious evidence for fit, for example, when speakers pick up cues from each other by repeating phrases, finishing sentences, or speaking in unison.

IV. THE INDIVIDUAL, AND THE SOCIAL CONSTRUCTION OF MATHEMATICS

"Nothing can replace a corridor for good scientific research."
Arnos A. Penzias, Vice President Research, Bell Labs[7]

Mathematical behavior is the activity of mathematicians. Research mathematics, as a socially constructed human activity, has a unique culture and tradition.[8] In this section, I look at the nature of this culture in order to examine how it is unique, and how this culture provides a basis for *inquiry math*. What must the student learn about this community, and about how mathematics is practiced in it? Piaget is correct, there is indeed an autonomy in the development of mathematical and logical abilities in the child,[9] and social influences "only play a part on condition that they are assimilated by the subject's structures" (Beth and Piaget [1966], p. 297). However, we have yet to explain the process whereby some students, in a matter of ten to fifteen years, construct a mathematics for themselves. And, from a constructivist perspective, we have yet to account for the fundamental role of the teacher in this process.

4.1. Mathematicians and the Myth of Proof

Classic descriptions of mathematical activity, such as those of Poincaré [1908] or Hadamard [1949], have focused on individual acts of creativity or insight. These accounts tend to single out isolated events as opposed to the activity itself. In some individual accounts, e.g. Hamilton's notebook [1843], we do get some insight into the logic of discovery,[10] but these are rare and typically only show an isolated mathematician. At best, these give a sense of the internal dialog, and this is the closest we get to a description of mathematical discourse. Thus, we have very few serious descriptions of mathematical activity.

Mathematicians, even more than scientists, are trapped in a logical positivist, ahistorical, view of their subject in which it grows "through a monotonous increase of the number of indubitably established theorems" (Lakatos [1976], p. 5). This view is preserved in official accounts of mathematics, namely *journal math* and *school math*. Both of these are based on a reconstructed logic that assumes that mathematics is a structured recapitulation of known results, and that mathematical activity is represented in the form of proofs. Self-conscious mathematics eventually turns to proofs as the archetypical mathematical activity. But proofs are only a small portion of the domain, and in general I see them as a backdrop to a discourse, not a part of it. Proofs are important in *journal math* and *school math*. They are the object of published papers and are referred to in presentations, but in the day to day discourse activities of a mathematician, proofs stay in the background.

Accounts of mathematical activity are thus held captive by what I will call a "myth of proof." This is a result of a glorification of proof as the activity of the mathematician and the only viable result of the activity of a mathematician. Before returning to the nature of mathematical activity, I must first provide an account of the notion of proof, and, in so doing, try to sharpen the distinction between the logic of discovery and reconstructed logic.

When we communicate a proof, it is common to say, "Let me reconstruct it for you." The reconstruction is an attempt to communicate an abstract object. We use the reconstructed logic of derivations to make proofs objective, in the sense of being repeatable, inter-subjectively testable, and not subjective. That is, insofar as we assure the objectivity of proofs it is through the repeatability of the derivation. Derivations are by definition in the realm of reconstructed logic.[11]

A derivation entices, or leads us into seeing a proof. It does not tell us (communicate directly) the proof. This is apparent in that we can often read, and reread, a derivation, and even commit it to memory, yet not see, in the sense of *understand*, the proof. Poincaré argues that understanding requires some additional act beyond an item-by-item comprehension.

> To understand the demonstration of a theorem, is that to examine successively each of the syllogisms composing it and to ascertain its correctness, its conformity to the rules of the game? For the majority [of people], no. Almost all are more exacting; they wish to know not merely whether all the syllogisms of a demonstration are correct, but why they link together in this order rather than another. In so far as to them they seem engendered by caprice and not by an intelligence always conscious of the end to be attained, they do not believe they understand. ([1908], p. 430–1)

Proofs then are the result of an act of "seeing" or an "aha." They too are removed from the account of the discovery, or better, they are a part of the meaning of the activity, but they do not provide insight into the logic of discovery. The question is not, what is the connection between concepts shown in the proof, but rather, how does one come to make the conjecture in the first place, or how does one then construct the proof, or how does one come to understand the proof. These are all questions that pertain to the logic of discovery. The reconstruction via the derivation is a different process. Moreover, as Lakatos argues, proofs themselves do not arrive ready-made on the scene. The meanings of mathematical concepts are negotiated, and are part of a process of conjecture, refutation, and modification that provides proofs with a history and a context. The actual history of a proof conflicts with the presentation of a proof as an ahistorical certainty that is typical of classroom presentations.

There is a distinct "pre-linguistic," or "non-linguistic" aspect to the logic of discovery and accounts often involve references to images or intuitions. Further, much of the "pre-proof" activity in which we engage is solely concerned with "picturing" a mathematical object and attempting to describe it. This activity is, by its very nature, imprecise. It is a start towards abstraction and generalization. It grows out of a "feeling" that there is (or ought to be) a connection between previously unrelated data. In spite of the "pre-linguistic" nature of the intuitions to which I refer, mathematical discourse spends a great deal of time attempting to convey descriptions. There is hand-waving, pencil or chalk sketches, and constant appeal to metaphors.

The communication of a proof only occurs indirectly, much as we would communicate the meaning of a poem, or a metaphor. G. Spencer Brown explains mathematical communication in a similar fashion.

> It is comparable with practical art forms like cookery, in which the taste of the cake, although literally indescribable can be conveyed to a reader in the form of a set of injunctions called a recipe. Music is a similar art form, the composer does not even attempt to describe the set of sounds he has in mind, much less the set of feelings occasioned through them, but writes down a set of commands which if they are obeyed by the reader, can result in reproduction, to the reader, of the composer's original experience. ([1969], p. 77)

The set of injunctions is the derivation. The taste of a cake, the sound of a symphony, or the awareness of a proof are all similarly indescribable. They are indescribable in the same sense as the elements of physiognomy that enable one to recognize an individual's face.

Proofs should lead someone else (or yourself) to make a construction. Proofs are social constructs and a product of mathematical discourse. The discourse itself, the activity of the research mathematician, is separate from the derivation (the *journal math* communication of the proof), or from the proof itself. The discourse of the mathematician, which is guided by the logic of discovery, produces mathematical insight and discovery. Thus, we use mathematical language to communicate on at least two different levels. In *journal math* (and *school math*) there is a direct attempt to lead another to the proof. In *research math* (and *inquiry math*) the proof enters only indirectly.

Mathematics itself does not face up to the distinction between proof and the actual activity of the mathematician. In presentations and papers one needs proofs. Problems in this perspective are obvious only when new types of mathematics are created. Gleick's [1987] popularized account of the introduction of the new mathematics of chaos provides a good sense of the problems with this assumption. When Mitchell Feigenbaum presented his new research, he was attacked directly on his methodology.

> Feigenbaum's work ... was not mathematics; he was not proving anything. He was studying numbers, yes, but numbers are to a mathematician what bags of coins are to an investment banker: nominally the stuff of his profession, but actually too gritty and particular to waste time on. Ideas are the real currency of mathematicians. (Gleick [1987], p. 178)

The role of proofs in presentations is made even more obvious in Gleick's account of the following exchange:

> Feigenbaum ... had barely begun to describe the [chaos] work when the eminent mathematician Mark Kac rose to ask: "Sir, do you mean to offer numerics or a proof?"
>
> "More than the one and less than the other," Feigenbaum replied. "Is it what any reasonable man would call a proof?" (Kac) (Gleick [1987], p. 183)

In fact, the audience accepted the presentation. But, I would argue, if Feigenbaum had not been making a presentation, and thus was supposed to be speaking *journal math*, there would have been little resistance to the argument.

As a result of the "myth of proof" the real activity of mathematics is hidden. Because of this myth, the better known accounts of this activity, e.g. Poincaré or Hadamard, focus solely on the "aha" experience – the individual's private act of insight or creativity. This single act, the making of a connection or the seeing of a proof, is a product of the activity. The activity itself is a process of inquiry, an ongoing dialog that one maintains with oneself and with one's colleagues. Students need to learn to continue this dialog with others and ultimately with themselves.

There is much more of the everyday behavior of mathematicians when we think of collaboration. It is very rare that mathematicians acknowledge the interaction so critical to their discipline. It takes someone like the Harvard mathematician Persi Diaconis, who was trained as a magician, to recognize the real dynamics of mathematics. He describes the importance of collaboration to the work of a mathematician. Moreover, he points to the type of discourse that is central to the activity of the mathematical community:

> Mathematical people enjoy talking to each other .... Collaboration forces you to work beyond your normal level .... Collaboration for me means enjoying talking and explaining, false starts, and the interaction of personalities. (Albers and Alexanderson [1985], as quoted in Schoenfeld [forthcoming], p. 27)

Mathematical people do indeed enjoy talking to each other. And the talk has a distinctly mathematical flavor. But mathematicians rarely put forth a proof. Rather there is conjecture, and imaging. And it is on this level that Polya's questions ring true. "Can you solve it another way?" "Have you solved another problem like this one?" These are the questions that mathematicians ask of others, and learn to ask of themselves herself.

4.2. The Development of Mathematics

> *A great discovery solves a great problem but there is a grain of discovery in the solution of any problem. Your problem may be modest; but if it challenges your curiosity and brings into play your inventive faculties, and if you solve it by your own means, you may experience the tension and enjoy the triumph of discovery.* (Polya [1971], p. v)

Mathematics changes because the people who do mathematics change. New or non-standard mathematics becomes standard because individuals in the mathematical research community, and ultimately the community at large, accept new ways of doing mathematics.[12] This process is analogous to the constructive process in the individual. The community (i.e., individuals in the community) must see the new problems as significant, and must accept the new methods and procedures. The old mathematics constrains what will be accepted by the community – but once accepted the new mathematics may transform the old. From a historical perspective, mathematics has undergone revolutions that radically transformed the nature, method, and problems of mathematics.

There are indeed lessons in the history of mathematics that may serve as catalysts for the development of mathematics in the individual. It is not the case that in some deep sense the child's development of mathematics necessarily recapitulates the historical development of mathematics ("ontogeny recapitulates phylogeny"). But, the curriculum itself follows an historical path. Many of the revolutionary changes that were adopted by the mathematical community throughout history are demanded of students. For example, the early nineteenth century rejection of number as quantity was a direct response to problems with imaginary and negative quantities. As students move from arithmetic to prealgebra they must come to grips with many of these issues.[13] It is not only for intellectual curiosity that students should "experience the tension and enjoy the triumph of discovery," as in the Polya quotation in the beginning of this section. Students need to learn the process of discovery because the very process of inquiry, which is so much a part of the history of mathematics and mathematicians, provides the mechanism for individuals to negotiate the revolutions for themselves.

4.3. Mathematics and the Individual Learner

In the course of a mathematical education, each student constructs some portion of what the mathematical research community has constructed over the past three thousand years. Individuals change because they accept new ways of doing mathematics. They see new problems as significant and accept new methods and procedures. Their old mathematics constrains what an individual accepts – but once accepted, the new mathematics may transform the old. From a cognitive perspective, individuals radically reconstruct mathematics as they learn, and this transforms the nature, method, and problems of their mathematics.

The process of becoming a member of a mathematically literate community involves a great deal more than competence, in the sense of obtaining only "the requisite skills, abilities and knowledge necessary for participation in a given community" (Mehan [1979], p. 3). Rather, as Mehan argues:

> ... competence is assembled by people in concert with each other. Therefore we must look to social situations, socially assembled situations, not individual persons as the units of analysis appropriate for the interactional display of interactional competence. (Mehan [1979], p. 4)

Becoming a member of a community requires the capacity to interact with members of the community – and this interaction is in the form of discourse. Becoming a mathematically literate adult requires a history of interactions with a community that speaks that language. Only through discourse and gradual interaction does a student participate in a consensual domain with a member of this community.

The traditional structure of classroom mathematics instruction is patterned after a faulty perception of mathematical activity. In the context of using the computer to help students making conjectures, Schwartz argues:

> There is something odd about the way we teach mathematics in our schools. We make little or no provision for students to play an active and generative role in learning mathematics and we teach mathematics as if we expected that students will never have occasion to invent new mathematics. We don't teach language that way. If we did, we would never require students to write an original piece of prose or poetry. (Schwartz et al. [1985], p. 1)

The introduction to mathematics that schools provide is far from the activity of the community of mathematically literate adults. I am not merely criticizing a discourse that begins with, "Did you turn in your homework?" More important, it is the very structure of most school discourse that points to its inadequacy for mathematical literacy. As can be seen in the initial dialog, the students are barely picking up a mathematical vocabulary. They have no idea of the proper syntax, or rhythm of a mathematical discussion, and no sense of making conjectures, or of evaluating mathematical arguments.

V. A SECOND DIALOG

Two days later we met for a second session. I began with the problem in figure 2.

$$n + 6n = 2n + 15$$

Figure 2. The Problem.

J: *What rules can we use?*

Teacher: *Every rule that you ever learned. Just do it any way you know how to do it.*

G: *Watch I'll do it and it will be wrong.*

Students: (Pause for 12 seconds, then laughter.)

Teacher: *What are you thinking about? Maybe we can talk about it while you are doing it.*

J: *Teacher said add negative of something, or do the reciprocal of 6, or I don't know which one.*

G:	*Multiply the reciprocal of 6 to both sides like that and that (*points at the screen*) and then do the reciprocal of 2.*
Teacher:	*If you just sort of look at the problem instead of trying to figure out how to solve it.*
G:	*Find out how many n's there are. Get all the n's together in one place.*
Teacher:	*OK, how would you do that?*
G:	*6n + 2n*
J:	*8n*
G:	*Plus n*
Students:	*9n*
Teacher:	*That's sort of true. What does the equal sign tell you?*
J:	*They equal each other. Both are the equation.*
Teacher:	*Okay, what is going on on the left hand side of it? Can you look at each side of the equation separately and ... what is going on in the left hand side of the equation? (Pause.) You have n + 6n. What would that be?*
J:	*n plus 6 times n. I don't know.*
G:	*6 times n to the 2nd power.*
Teacher:	*Part of it is if you have 6 n's, I mean think of it this way if you ...*
G:	*7 n's.*
Teacher:	*Exactly, do you see that?*
J:	*I think*
G:	*6n + 1n is going to be 7n.*
Teacher:	*Okay, now you have 2n + 15 and the problem is you want to get the 2n over to the left hand side, right? In other words you want to get all the n's ...*
J:	*So you add a negative two. I don't know.*
G:	*So you ... like you're gonna times the reciprocal of that to this side. No?*
Teacher:	*What would that be?*
J:	*Don't you add the opposite and a negative 15?*

From a content perspective this is a disaster. Moreover, the students are still struggling, and very insecure, from G's "Watch I'll get this wrong," to some very meaningful, "I don't knows." But from the perspective of participation there are several very encouraging aspects where the two are cooperating – picking up on what the other is saying. Thus J's comment "Teacher said..." is a passive attempt to report, but G picks up on it and tries to explain. This was not an optimal solution, and I tried to get them to reflect on the problem. G then provides the first sub goal, "Find out how many n's there are." The two then work together, ultimately ignoring the equal sign and getting "9 n's." The key here is that they did work together. G and J fall into a rhythm. G says "6n plus 2n" and J fills in "8n". G adds "Plus n" and together they say "9n." Right or wrong, this part of the episode needs to be noticed because it has a very different quality.

The difference between the two dialogs depends on the fact that in the second dialog the participants arc beginning to "turn around" together. They are beginning to create a consensual domain.

VI. THE SCHOOL AS SOCIALIZING FORCE FOR MATHEMATICS

In order to work within the mathematical community one begins by accepting the methods and problems of the contemporary mathematics.[14] As Kuhn [1963] points out,

> Scientific education inculcates what the scientific community had previously with difficulty gained – a deep commitment to a particular way of viewing the world and of practicing science in it. That commitment can be, and from time to time is, replaced by another, but it cannot be merely given up. (Kuhn [1963], p. 358)

The "particular way of viewing the world" is the main product of a graduate education. This is in sharp contrast to the particular way of viewing the world that is engendered by the school mathematics curriculum. The process of graduate education brings the student into a particular research community with its own methodology and research program, and its own discourse. There is a continuous process of acculturation that gradually introduces the student to the creative processes of the community. This is fostered by smaller classes and seminars. Perhaps even more important are the informal conversations with faculty and other graduate students.

We can now turn our attention to the difference between *school math* and *inquiry math*. It is my contention that the very structure of *school math* discourse is antithetical to the nature of mathematics as a creative discipline. *School math* presents the subject as a collection of facts and procedures. The task is fundamentally structured as an information transfer. The teacher tells the class about a procedure. The teacher then asks a question. There is a single correct answer, already known by the teacher. A student is selected by the teacher to answer the question. The teacher then evaluates the answer.[15]

Classroom dynamics are also managed by a teacher to minimize teacher-student and student-student interaction. This can be seen in the wait-time experiments (cf. Rowe [1974], and also in recent research as reported by Carlsen ([1989]). Teachers control discussion through a variety of subtle, and some not so subtle cues; by establishing classroom routines; by stringing together sentences with words like "and," "but," and "so;" and immediately answering their own questions (cf. Carlsen, p. 7 *ff.*).

As I said before, there is a consensual domain established in *school math* classrooms. Teachers and students "cooperatively contribute to the social organization of the classroom" (Mehan [1979], p. 27).

The synchronization of students' replies with teachers' initiations together with the integration of interactional form with academic content are constituent features of participation in the classroom community .... When the teacher initiates a round of questioning, students are expected to produce academically correct replies in ways that are consistent with the normative expectations associated with the turn allocation procedures operating in the classroom at the moment. (Mehan [1979], p. 16)

The nature of the discourse in this consensual domain is very restricted. Students do not carry on conversations together, and in fact never address one another. Instead, all dialog originates with the teacher who remains in control, and speaks every other turn.[16] As Lakatos argues, "It has not yet been sufficiently realized that present mathematical and scientific education is a hotbed of authoritarianism and is the worst enemy of independent and critical thought" ([1976], p. 143).

It is therefore no surprise that Alan Schoenfeld [forthcoming] attacks the dichotomy between mathematics and sense-making that permeates the general culture.

What we see here ... is the totalitarian power of mathematics. Throw enough formalism or appeal to mathematical authority in front of people, and they'll back down. For some reason people don't expect heavily mathematical statements to make sense to them. (Schoenfeld [forthcoming], p. 7)

Throughout this paper Schoenfeld points to examples where the tyranny of mathematics numbs people into submission. Schoenfeld argues that "such suspension of sense-making develops in school, as a result of schooling" (p. 10). Indeed, it is in the very nature of *school math* discourse that students are learning to suspend the requirement that mathematics make sense.

School provides the student's main experience with mathematics, and this is markedly limited. The mathematics teacher ought to be the socializing force helping the student become a mathematically literate adult. However, the dynamics of *school math* discourse is based on an entirely different understanding of the nature of learning and mathematics. The school context contributes only minimally to a student becoming mathematically literate. It does not encourage the student to question, to challenge, and thus to learn anything about real mathematical behavior. In contrast, inquiry-based activities provide ownership of ideas, and allow the student to construct mathematics.

Mathematical discussion is a unique feature of a mathematically literate adult. While this is a mainstay of graduate education, it is possible to participate with modest mathematical training and experience. But it is important to emphasize that the experience must be as a member of a mathematical culture, and the recitation of facts, or the memorizing of procedures is a very different activity. While a teacher can serve as a moderator or facilitator through parts of the discussion, there is a need for the teacher to play the role of a genuine participant.

In a consensual domain in which communication occurs, each participant must respond to the other. The changes that are required to adjust the participants to each other entail that each participant in the domain must be prepared to learn. This is a real indicator that the nature of the discussion is genuine. Confrey reports,

> Through the process of the interview, my own conception of exponential functions was transformed, elucidated and enriched. Our minds are such sensitive instruments and as we strive to examine our understandings, the character of those understandings will inevitably shift; this is the essence of the process of reflection and construction. ([1988], p. 8)

Moreover, in my judgment, the fact that both the teacher and the student are prepared to learn is the essence of good teaching, because the teacher is placed in a position to listen and participate in a discussion. Students and teacher can jointly pose questions, and jointly set out to answer them. The benefits of this mutual participation are clear. As Lampert notes regarding an episode in her class, "Since there was a disagreement in the class about the meaning of .0089 this problem and its resolution belonged to the students. This is a rare and valuable turn of events in the school classroom, where most problems belong to teachers or textbooks" (Lampert [1987], p. 25).

A genuine mathematical discussion does not just happen – it evolves, and the members of the discussion have a commitment to follow the discussion. The dialectic of conjectures, proofs, and refutations creates the meaning of a concept.[17] Through this constructive process the individuals each create the mathematics anew in the course of the discussion. It is this creative energy that is missing from school mathematics. By emphasizing the individual's need to create mathematics, the constructivist perspective provides a basis for restoring this energy to the classroom. The participants utter speech acts of belief, certitude, acceptance, rejection, and questioning. These have a very different character from the declarative speech of *school math*.

VII. PROBLEM SOLVING – LESSONS FOR DISCOURSE

> *What we are asking for is a methodological and psychological shift in emphasis in the role of teacher and student, a fundamental change in the nature of the classroom environment.* (Postman and Weingartner [1969], p. 205)

In this section, I examine three of the more promising proposals for revising the mathematics classroom. Each of these adopts different strategies to circumvent the standard teacher-controlled *school math* classroom. These approaches are representative of a variety of proposals that contain specific strategies for improving teaching, and ultimately for improving a student's performance. It is clear that many of these approaches are effective. In articles describing them, authors spend a significant effort describing the type of discourse, the role of the teacher and students in the discourse, and clear instructions for carrying on this discourse. But this is seen as secondary to the main thrust of their arguments. The proposals are interpreted in various ways by the authors, for pair problem solving, for modeling and executive decisions, for reciprocal teaching, and so on. With the exception of pair problem solving, these approaches have not spread much beyond the particular research environments in which they were developed. Perhaps some of the difficulty in dissemination is due to the ambiguous way in which the approaches are explained. On the one hand, recommendations are made for revising the quality of classroom discourse. On the other hand, claims are made regarding student's thought processes, or problem solving rules and dynamics. From a constructivist perspective, we don't have direct access to the latter, which is why we make models. Referring to this realm misleads both researchers and their readers and changes the focus of the descriptions.

So far I have argued that discourse is the central feature of teaching math inquiry. In this section, I argue that each of these proposals is better understood – and implemented – from a discourse perspective. The net effect of each of these proposals is to revise classroom discourse. When viewed from this perspective, they contain a number of specific suggestions for how to change the discourse in a classroom.

7.1. What Words to Use

An articulation of mathematical discourse is a central but subtle part of the problem solving literature. Polya begins his heuristics studies with a list of questions and suggestions which represent a dialog that he wants the problem solver to carry on.

> If, using them properly, you address these questions and suggestions to yourself, they may help you to solve your problem. If using them properly, you address the same questions and suggestions to one of your students, you may help him to solve his problem. (Polya [1971], p. xix)

Polya's intent in this statement is pretty clear. The questions that arise, and the manner in which they are presented, are instructions for carrying on a conversation. Used in this manner, they are specific discourse rules for a teacher. Viewed as discourse, Polya is describing the activity of successful mathematicians and inviting others to participate in this domain. The discourse aspect of Polya's work is often ignored because of his own ambiguity regarding his task. Thus, at a later point in the book, he states that, "Modern heuristics endeavors to understand the process of solving problems, especially the mental operations typically useful in this process" (pp. 129–30). He then maintains, "Our list is, in fact, a list of mental operations typically useful in solving problems" (p. 130). This reinterpretation of the list has confusing epistemological and pedagogical implications. Epistemologically, we have no grounds for this sort of claim. Moreover, viewed pedagogically, he is presented as offering a series of cognitive injunctions or rules that one is supposed to memorize and follow. This distinction is important from a constructivist perspective.

In our terms, Polya suggests that the student and teacher construct a consensual domain that is open to mutual participation, and that fosters a conversation. If the teacher is careful to speak in a certain way, and to ask questions ("Do you know a related problem?"), then the student, by participating in the domain, learns this language, learns to solve problems, and begins to act mathematically. This is much more in line with the spirit of Polya's suggestions than asking the student to memorize a set of rules that can be applied while solving problems (for example, "Is 'Find a related problem,' Rule 3?").

7.2. *The Math Teacher as a Model*

Alan Schoenfeld's contemporary proposal is closest to Polya's. In many ways, Schoenfeld's self conscious accounts of his own classroom experiences take Polya a step further. He argues, "Mathematics problem-solving instruction to date has focused largely, and with questionable success, on heuristics or tactics" ([1983b] p. 346). The problem is that teachers are telling students about problem solving or about heuristics or tactics. Schoenfeld comments, "Such lectures by themselves are almost guaranteed to have no effect on the students' behavior ..." ([1983a], p. 29). Instead, he sees a need for direct intervention while students are solving problems. Again:

> Students learn by doing, not by watching .... Lectures are kept to a minimum: even when there is a point that I want to get across, it can be made much more powerfully if the students have grappled unsuccessfully with a problem before they are shown why another type of approach "makes sense." ([1983a], p. 26)

For Schoenfeld, "... the primary responsibility of mathematics faculty is to teach their students to *think*: to question and to probe, to get to the mathematical heart of the matter, to be able to employ ideas rather than simply to regurgitate them" ([1983a], p. 2). Schoenfeld suggests that a teacher serve as a role model in three ways. First, the teacher may take the time to solve a homework problem in front of the class, talking out loud about the various options. For Schoenfeld, this is helpful, but the presentation is one-way and the student needs to be an active participant. Second, and more important, the teacher is to serve as a "moderator," an "orchestrator of ideas," and an "alter ego that raises important questions and keeps things on track" ([1983a], p. 13). The teacher is literally playing the role of a dialog coach. As Schoenfeld argues, the questions that he asks eventually become second nature to the students, "By the middle of the semester I can ask them, 'OK, what question am I going to ask now?' and they can usually tell me" (p. 13). Third, the teacher models good problem solving behavior by challenging students to bring in problems for the teacher to solve, and talking aloud through this process. This puts the teacher in a genuine problem solving situation during which students can observe how the teacher avoids pitfalls, recovers from wrong moves and missteps, and genuinely uses the techniques that are the focus of the rest of the course. From my own experience, this full participation of the teacher, with the potential to make errors and to fail is very important. You can't fake mistakes, or feign a discovery.

In spite of these valuable insights, the role of discussion seems to be secondary in the articles. This is especially important in light of the fact that in Schoenfeld's problem solving course eighty-five per cent of the time is spent in class discussion, with the bulk of this time spent in small group discussion.[18] Schoenfeld circulates around the class, observing and interacting, playing the role of a discourse coach. This aspect of his teaching, the role of discourse coach, is crucial to the success of his course and is also the most difficult aspect to convey. Moreover, the classroom dynamics, which is the focus of this essay, does not appear explicitly in published accounts. As he explains, "... the classroom style did not reflect the 'revealed truth' exposition of this vignette. Instead, the classroom dynamics reflected the dynamics of real mathematical exploration" (Schoenfeld [forthcoming], p. 37).

On several occasions Schoenfeld provides what he terms "distilled versions" of classroom discussions. These provide some insight into the process, but are mostly driven by the mathematics. These versions show the false starts, the ebb and flow, and then the discovery. Interestingly, from our perspective, in a four-page presentation of a class discussion there is only one mention of the process, which appears in a footnote. Schoenfeld's comment is, "The 'planning decision' summarized on the previous page took about five minutes of actual discussion in the classroom. I played the role of moderator, asking questions like, 'All right, what choices do we have?'" ([1983a], p. 45). While a reader can get a feel for the sort of culture that is being established, it is hard to transfer this dynamic to one's own teaching without more explanations of the discourse. Schoenfeld's account focuses on the mathematics. However, as he acknowledges, this is not what is unique about his classroom.

In Schoenfeld [forthcoming] there is a second distilled version of a classroom discussion. Again, Schoenfeld focuses on the content of the discussion, not the process. I want to look at how one can carry on this sort of dialog. It is clear that the underlying consensual domain of this class is very different from *school math.* In this case, the problem is to fill a 3 x 3 magic square with the digits 1 through 9. In his account, Schoenfeld shows how the class was engaged in a principled solution to the problem. He provides a Polya-style commentary on the heuristic principles guiding this session. To this point the session seems enlightened, but structurally similar to many other classroom sessions. The teacher knew what the answers were, and could easily predict the flow of the discussion. No serious mathematician would be interested in being "the student." It is at this point that Schoenfeld becomes a genuine participant in the discussion and the goal of the session clearly changes. It is not to *get the answer.* "the goal is to understand the magic square" ([forthcoming], p. 34). Can we solve the problem a different way? And more important, "What about extensions? Generalizations? Up to this problem we've only answered someone else's problem. We make it our own by playing with the ideas it raises" (p. 35). These are the words of a mathematician. This is the beginning of an *inquiry math* discussion with the class.

Why does this session work? Partly because Schoenfeld is asking the right sort of questions. More important, because he has established a domain of discourse with the students in which they have come to expect the questions that he would ask. It is not just that these are questions in the Polya mold: "Can you ask a more general question?" but rather, these questions are being asked in a broader context. The quality of the discourse makes them legitimate questions to ask. The goal of the discourse is for the students to understand and, in the process, make the problems their own. Even though the questions may appear general, the participants are asking mathematically relevant questions. The questions are based on the particular mathematical context.

In the second part of the classroom sessions, Schoenfeld models a mathematically literate adult talking to other mathematically literate adults. The effect of this type of modeling is very powerful. In one of my own sessions, J had been doing very well and G could not catch on. I was waiting for a while, giving G some time to solve the problem, when J turned to her and assumed the role of an interlocutor: "What are you trying to do?" "What is the major operator?" "How can you get rid of it?" I was genuinely surprised to see J mimicking my own teaching style – not merely in the choice of words, but also in the tone and the role she assumed. She was passive, not judgmental, and very positive, even when G made some incorrect suggestions. I postulated that J had internalized some of this for herself, and that this internalization provided a ground for the success she was experiencing at that time.

Teachers with a weak mathematical background might feel intimidated by the mathematical content of an open inquiry-oriented classroom. Moreover, mathematicians are usually no better than other teachers in conversing with their students. Although they may have learned *inquiry math* in their studies, they revert to *school math* in the classroom. Schoenfeld is rare because he naturally speaks *inquiry math* to his classes. It is certainly possible to argue that Schoenfeld's success is idiosyncratic, and due to his own special combination of mathematical prowess, and interest and ability in teaching. I would argue that, while some subject matter expertise is important, it was Schoenfeld's willingness to explore that was much more important to the success of the class. It is clear that Schoenfeld's understanding of the mathematical content provides a depth and flexibility to his classroom. Nevertheless, the central feature of the classroom culture is his language. In the next section, I look at a case where mathematical sophistication does not play a central role.

7.3. Pair Problem Solving – Experience With Math-Talk

The effectiveness of talking *inquiry math* with non-mathematicians is explored in pair problem solving (Whimbey and Lochhead [1980], and Lochhead [1985]). Pair problem solving was implemented as an overt attempt to make students active learners. This approach directly challenges the role of the teacher as "information provider and answer verifier." Instead students share this role.[19] The central feature is that students alternate roles in the problem solving process. In the pair, each student takes turns as *problem solver* and *listener*.

In addition to the theoretical basis in active learning, there is explicit attention to the quality of the dialog. Whimbey and Lochhead refer to their teaching style as Socratic or "teaching according to the Talmudic tradition." Moreover, they spend a great deal of time describing the nature of the roles the students are playing. Lochhead's [1985] explanation of the process emphasizes the centrality of the role of the listener. One student asks the other a question, and listens to the answer. In Whimbey and Lochhead [1980] there is a detailed explanation of the listener's role. These are, quite explicitly, the rules for being a good listener, and for carrying on a conversation of a very particular sort. The listener's job is to check for accuracy, and to demand vocalization on the part of the problem solver. The early problems, which are very simple, are designed to allow a conversation to begin.

From my perspective, the success of this approach is due to the fact that conversation is forced on the student. In the act of trying to explain to another student (the listener) the decisions and the specific moves in a solution, the problem solver is learning to verbalize (consider, put into language) their thinking. This requires reflection, but it also means that both students are engaging in *inquiry math* discourse. In order to understand this dichotomy of the description of the success in terms of active learning, and the attention paid to the quality of the discussion, it is important to emphasize that pair problem solving evolved out of very different considerations.

Pair problem solving grew out of the think-aloud strategies used in cognitive science. These strategies were adopted because the process of problem solving was, by definition, observable only by the person doing the activity. If subjects could talk about what they were doing while they were doing it, then an observer might gain some insight into their operations. In a teaching situation, it seemed important to show students how an expert solves a problem. Conversely, a teacher listening to a student might gather some clues as to where the student was falling short. But, in a Heisenberg-type switch, the very act of talking aloud required a level of reflection that aided in the problem solving process and ultimately had the power to change the student's performance. This is the first unexpected result of changing the standard mode of interaction in a classroom.

Because this process is time consuming for a single teacher, the key step for Lochhead and Whimbey was to use another student as a listener, providing her with the answer, and letting her play the role of teacher in the interaction. The students alternated so that both could have a chance to solve problems. This had a second unexpected result when it became clear that students were learning in both roles.

In the terms of this paper, the students created a new sort of consensual domain in which rules were established that allowed both participants to play a role in a mathematical conversation. Each was learning to talk and to listen in a mathematical context. Their interactions were mutually supportive, encouraging them to provide hints and to help in other ways. The teacher was not a teacher of mathematics, but a dialog coach. "Later, as [the students] skills improve, the teacher provides feedback on how well individual students are performing in the role of listener or verbalizer, with advice on how to do better" (Lochhead [1985], p. 112).

VIII. CREATING THE CONTEXT FOR A DISCUSSION

In class, try to avoid telling your students any answers .... Do not prepare a lesson plan. Instead, confront your students with some sort of problem which might interest them. Then, allow them to work the problem through without your advice or counsel. Your talk should consist of questions directed to particular students, based on remarks made by those students. If a student asks you a question, tell him that you don't know the answer, even if you do. Don't be frightened by the long stretches of silence that might occur. Silence may mean that the students are thinking. (Postman and Weingartner [1969], p. 194)

The process of establishing a culture for inquiry requires that the classroom allow conversations between teacher and students, and amongst students, and that these be discussions – conversations in which communication occurs. For a discussion to occur in a classroom, there has to be a consensual domain that supports significant participation on the part of students.

The main effects of each of the specific proposals above, and a host of similar ones are, first, to change the classroom culture by creating such a consensual domain and, second, to change the classroom discourse to allow for discussions. By focusing on these effects, the different proposals can be seen to share a common objective. It is then possible to suggest guidelines for evaluating the effectiveness of the different proposals in achieving this objective. These proposals are designed to create inquiry-oriented classrooms, in the sense that the students are engaged in inquiry – that is, active investigation of a problem.

This is not to say that these different strategies produce identical classroom cultures. But, there are some common features: in each case, the teacher is no longer functioning as the main provider of information; the locus of control has, to some extent, shifted from the teacher to the students; and, in many ways, there is a deliberate undermining of the authority of the teacher. These classrooms are significantly different from pure "discovery" classrooms, in which the teacher provides an environment for students to explore and discover for themselves solely through the structure of the environment. Rather, these classrooms fall in the category of what we have called "inquiry by design" (Feurzeig, Richards, and Roberts [1987], pp. 2 *ff.*)

Inquiry by design, has three distinct aspects. The first concerns the role of the instructor as designer: to motivate the student to think actively about mathematical issues. Polya's first principle of teaching is to make the student an active learner (Polya, 1962). Teachers create an inquiry environment by design and careful planning. A consensual domain that fosters inquiry is not generated simply by removing structure, but rather by creating a different structure, one that provides support for the student to participate. It is necessary to provide a structure and a set of plans that support the development of *informed* exploration and *reflective* inquiry without taking initiative or control away from the student. The teacher must design tasks and projects that stimulate students to ask questions, pose problems, and set goals. Students will not become active learners by accident, but *by design*, through the use of the plans that we structure to guide exploration and inquiry.[20]

The second aspect of *inquiry by design* is that the student must learn to inquire systematically, that is, the student must learn how to inquire. Polya (and Schoenfeld) provides the student with a design or structure for their inquiry. The students learn to approach a problem with a plan in mind and to ask specific questions of themselves when they are stuck. There is a method to their inquiry – a design.

The third and most important aspect of *inquiry by design* identifies the student as designer. Learners must actively construct their own knowledge. This focuses on the constructive aspect of design as an essential complement of analysis in a student's progressive construction of mathematical knowledge. Our experience is that the activity of designing and building mathematical objects and processes can be used to motivate students' understanding of mathematical structures and the sense and methods of mathematical inquiry.[21]

An inquiry environment provides the context for investigation. Mathematical inquiry requires further that there be mathematical discussions. These are, paradigmatically, discussions between mathematicians. This is epitomized in the *research math* language, described earlier. As David Bloor argues:

> Clearly an aura, a certain feeling, surrounds the characteristic patterns which exemplify mathematical moves and this aura can now be identified as a social aura. It is the effort and work of institutionalization that infuses a special element and sets apart certain ways of ordering, sorting, and arranging objects. (Bloor [1976], p. 88)

Mathematical inquiry can occur when the participants are actively engaged in solving a problem in a mathematical domain. Inquiry discussions generally share much with the social aura of mathematical discussions. A mathematical discussion occurs in a mathematical context, with the participants asking mathematical questions.

In order to create a consensual domain for inquiry discussions, it is necessary to maintain a consistent level of discourse. Small but consistent changes in classroom dynamics have significant effects on the quality and content of classroom discourse. As I mentioned earlier wait-time research, which examines the amount of time a teacher pauses after asking a question, has demonstrated these effects dramatically. As Carlsen points out, in his recent survey of question-asking in the classroom, the most interesting findings of the wait-time research is that "the cognitive level of teacher questions rose when teachers were trained to increase wait-time" ([1989] p. 16). In extended wait-time classes, teachers talk less and interrupt students less often ([1989] pp. 16–7). "Extending wait-time also has an effect on students' verbal participation in classes. Among the effects reported are an increase in the number of student utterances, the length of student utterances and the complexity and cognitive level of student responses" ([1989] p. 17). In terms of our earlier discussion, this changes the rhythm of the discourse. By extending the wait-time the teacher assumes the role of listener. This is obviously a serious change, and a role that the teacher needs to learn.

Lochhead's presentation of the task of listening for pair problem solving provides some direct suggestions for the teacher's preparation for beginning a discussion. This role-change alone significantly alters the dynamics of the classroom. If, as Lampert ([1986], p. 339) argues, the teacher's role is to build a culture of sense-making in the classroom, then the teacher must make an effort to listen and to understand the student's perspective and questions. As Lampert comments about an episode in her class, ([1987], p. 24) "I needed to figure out how whatever they were thinking about decimals might be seen to make sense, rather than proceeding as if they were on the wrong track."

The teacher must first strive to make the classroom safe for all the students to participate. This is an important notion that draws from the brainstorming literature on how to conduct a productive meeting. A good leader:

> ... sets top priority on the defense of each person's image of himself. He knows that each member cherishes his own individuality above any problem to be solved. If this quality is threatened, the member not only stops cooperating, he becomes dangerous to the purposes of the meeting. To prevent this the leader repeatedly demonstrates that in this meeting no images will be damaged, no one is going to lose. (Prince [1970], p. 5)

This requires that the leader adopt several strategies throughout the discussion:[22]

- Be a 200 percent listener to your team members.
- Do not permit anyone to be put on the defensive.
- Use every member of the team.
- Do not manipulate your team.

These carry over immediately to a classroom discussion. Moreover, Prince's theory of running a meeting requires that the team members switch roles and alternate as leaders, thus anticipating the reciprocal teaching strategy of

Palincsar and Brown.

In traditional teacher-centered classrooms the initiation-reply-evaluation discourse format results in a highly structured consensual domain. This provides a sense of security for the teacher. The teacher knows the questions that are to be asked, and is provided with the answers. The teacher moves from item to item in a lesson plan (usually supplied with the text). The rate of progress is based on the teacher's evaluation of student responses, together with a consideration of upcoming deadlines, usually in the form of pre-scheduled testing periods.

In more open-ended classrooms, teachers are less secure, their authority is more apt to be challenged, and they may not know the answer to a question. Students are also less secure, and are not sure what is expected of them. They are not convinced the new structure (or lack of structure) will be sustained, and they are not sure how they will be graded. This places a much heavier burden on both the teacher and the students. Teachers must feel secure in their knowledge of the subject area, and as noted in this section's opening quotation, they must have the courage to admit they do not know an answer. This provides a genuine opportunity for students to see what the teacher does about it, and equally important, the teacher's admission gives students permission to make mistakes and to face their own ignorance.

Proper teacher preparation is essential for the success of an inquiry classroom. It is important to reiterate that inquiry by design requires a good deal of structure; that teachers create an inquiry environmental by design and careful planning; and that teachers need to feel comfortable with the subject matter. Teachers must design tasks and projects that stimulate students to ask questions, pose problems, and set goals. But all of this is futile if they do not create a classroom culture that supports these activities. The teachers need to know the mathematics, and they need to know how to conduct inquiry discussions.

Inquiry-oriented classrooms are perhaps best distinguished by, what might seem an accidental characteristic, an intensive pursuit of a particular topic or question over a period of time. Typically, the teacher provides, what we will call, a *setting* that persists for several class sessions, with an entire class session often devoted to resolving a single question.

By taking the text as our basic unit of meaning, this paper provides a principled reason for organizing classes around settings. Discussions take place within a context and require a common language and a common understanding of underlying assumptions and open problems. A setting that remains common for several weeks provides a basis for a discussion. It literally provides a *context* – a semantic and pragmatic frame.

Settings can be abstracted from real world situations, such as running a store, making a scale drawing of a room, or organizing an airline schedule; or they can be entirely fabricated to suit subject areas and topics. In contrast with Postman and Weingartner, a setting does not have to be "relevant" to a student's life. Nor, in contrast with proponents of "situated learning," does a setting have to draw from realistic situations. If you can find a situation that is relevant and realistic then it is almost always good, but neither is necessary. The central feature is that the setting be rich enough and engaging enough to allow exploration for several weeks.

The notion of a setting draws from suggestions of Dewey and Piaget that work be organized around projects, rather than problems. A key feature is that the class be permitted to fully explore a single issue. This needs both an attitude on the part of the teacher and careful selection of an information rich setting. In the course of exploring a simple problem, filling in a magic square, Schoenfeld generated a host of new, generalized, problems and proceeded to create what appears to be a week's worth of curriculum. Lampert's inquiry classes are also organized around large problems that students can't solve immediately, or by a formula. The single problem provides a context in which to carry on a discussion. This allows students to become familiar with the issues, and to develop a vocabulary that is stable for the problem.

As we saw in the episodes quoted earlier, inquiry discussions do not happen quickly, nor do they happen simply because the teacher wants them to occur. Establishing the new consensual domain for a classroom may take months. Sessions in the beginning may feel awkward for all involved. They are awkward because the rules are being explicitly changed, and we are not used to this process. A significant amount of time must be spent teaching the rules for operating in the domain. Changes in the domain need to be initiated carefully, with a good deal of support for the student. The teacher's expectations should be limited in the beginning. Students need to feel free to participate in a way that allows them to make mistakes in public. This makes students vulnerable, and the classroom culture has to protect them.

In-depth questions that are worth an entire class session require first that time has been allocated for students to learn the vocabulary and the rules of the inquiry game. This is a staged process similar to the notion of scaffolding (cf. Bruner, and more recently, Palincsar and Brown [1984]). There are three stages:

1. Etudes. The first stage is designed to introduce the student to the setting. The teacher provides a good deal of structure to acquaint students both with the language and problems of the setting, and with the tools available. This stage can tend to be fairly pedantic and may involve more emphasis on computation than later stages, but it provides a base line of content about which to communicate. The name etude is derived from piano finger exercises that, although potentially interesting in and of themselves, are fundamentally preparatory.[23]

2. Explorations. The second stage is designed to provide some structure for the student in the explorations of the setting. The teacher suggests certain goals for the students that open up some of the possibilities of the setting. These goals may be open-ended, but the teacher is still largely influencing the direction of the exploration. Gradually, the problems change and are more exploratory in nature. Finally, the students are in a position to explore on their own.

3. Excursions. In this final stage, students play an active part in determining

the direction of the inquiry. A key feature is that students pose their own questions to pursue. Consequently, different students, or groups of students, may pursue different problems.

The problems inherent in this approach are obvious in the episodes with the *Algebra Workbench*. The *Workbench* provides a restricted setting and, in theory, could fit into a fairly traditional classroom. To begin with, students are comfortable with classroom problems consisting of solving equations. The equations are not placed in a wider context, and there is no attempt to provide interpretations for the variables. The *Workbench* is in two parts. An algebra calculator helps students solve equations by performing computations (multiply both sides by 'n + 1'), and an expert system that can solve linear equations, is able to provide advice or do a step. The *Workbench* is an example of a non-prescriptive tutor because it remains under the control of the student. Computations are performed, advice is proffered and steps are done only upon the student's request.

IX. THE WORKBENCH REVISITED

In the earlier two dialogs, the students were struggling to communicate. In subsequent sessions I tried several different strategies to help them talk to me, and to each other, about mathematical issues. They took turns solving a problem. They assumed different roles – one controlling the mouse, the other telling what to do; one explaining how the other had solved a problem; and, one making up problems for the other to solve. My initial goal was to get G and J to interact while solving problems. This would be the basis for establishing a new consensual domain, something different from the initiation-reply-evaluation domain from school math.

In this section we look at two brief vignettes from a session two weeks later. In these vignettes we see glimpses of communication within this new domain. These vignettes are not ideal interactions, rather they show a clear mix of success and failure. A teacher's most difficult task is to change their goals, and most important, their understanding of how school works. This may be counter-productive in their other classes!

Vignette 1. Using the *Workbench*, G makes up a problem for J:

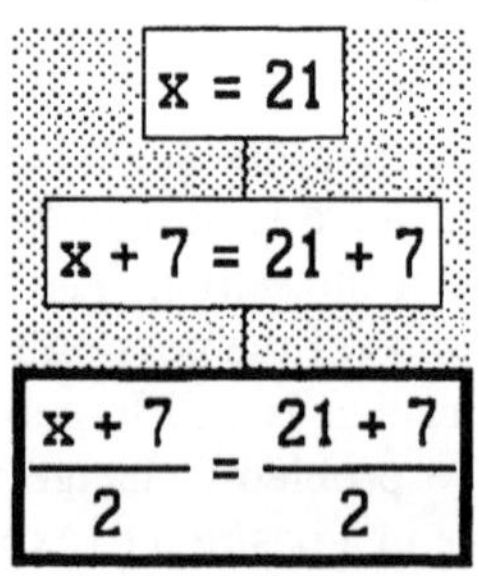

Figure 3. G starts with x = 21, adds 7 to both sides, and then divides both sides by 2.

J: *(Trying to solve the problem, chooses Do Math, which adds 21+7, multiplies both sides by 2, subtracts 7, and chooses Do Math, which subtracts 28–7.)*

G: *I told you it was easy.*

Teacher: *Excellent. Now go up to the top. Right about there, that's good enough. Click on the third box. Click right there. Now, I could do that, I could solve this in two steps.*

G: *Sure.*

Teacher: *Can you?*

G: *Sure.*

Teacher: *What would you do?* (G selects 21 + 7 and Do Math. See figure 4) *That's what I did.*

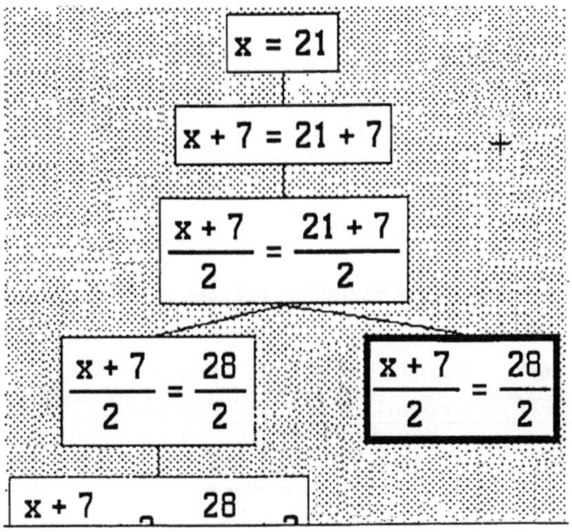

Figure 4. G tries the same stategy as J.

G: *How about this whole thing.* (Selects right side.) *That's one step.* (Chooses Do Math. Nothing changes. Undoes step.) *No, it's not one step. Go ahead, what would you do?*

Teacher: *How did you get …. Explain to me what you did to get to that problem. You made it, so I want you to start at the top and tell me how you got it.*

G: *What do you mean, got it?*

J: *What did you do. Added 7.*

G: *Oh, I see, so I am going to …* (Highlights 7, pulls down **Both sides, subtract** and pauses.)

Teacher: *Try it* (See Figure 5.)
G: (Highlights 2.) *And then I add it.*
29 *J:* *Multiply.* (See Figure 6. Pause.)
Teacher: *That got very ugly. What...* (The session continues.)

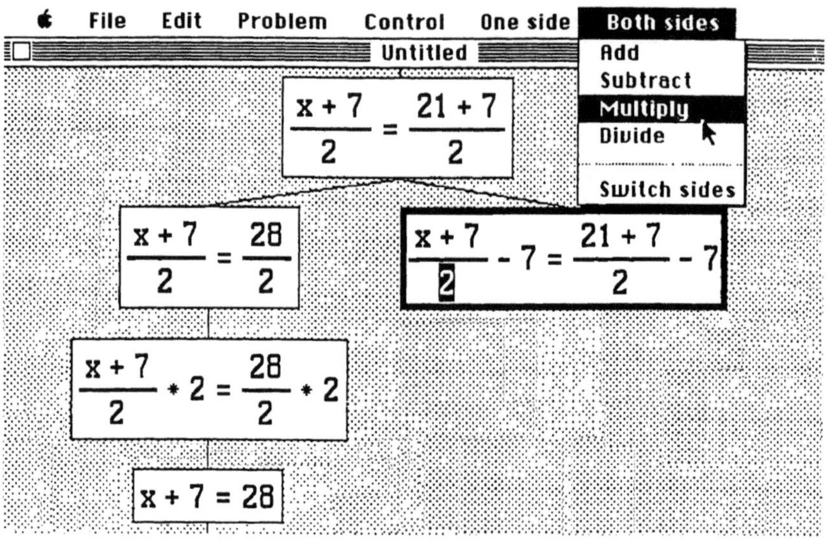

Figure 5. G subtracts 7 from both sides, and goes to multiply both sides by 2.

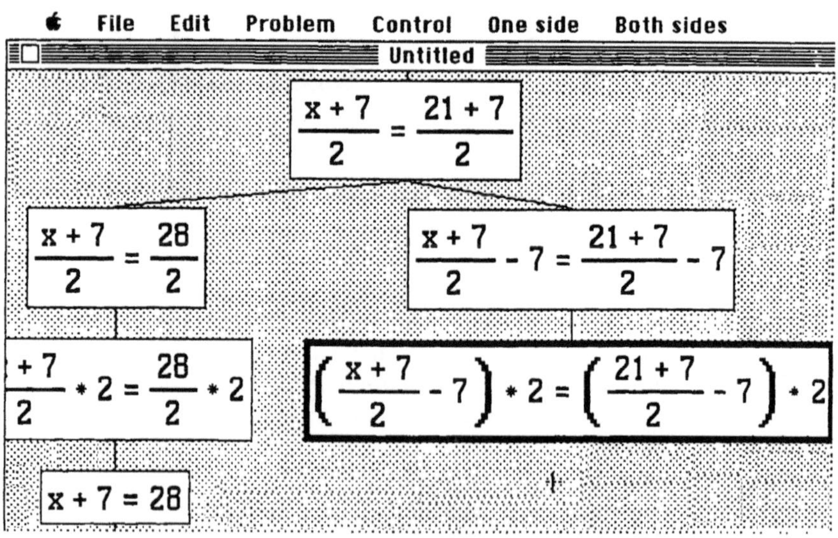

Figure 6. The problem gets complicated.

This session was designed to get the students to think about altenative ways to solve a problem. In particular, G had taken two steps to make the problem, first adding and then dividing. The easy solution would have been to reverse the operations, viz. to multiply and then subtract. In general, both G and J had a tendency to do the "easy" operations first. For example, they both started by adding 21 and 7. When I challenged G (09), there was an easy interchange and she tried a couple of steps that didn't work.

G saw pretty quickly that she needed to use the converse of the operations (15–20). In the continuation of the episode (not included here), G took some time before she understood that she also had to reverse the order of the operations. After doing this, she said, "*Hey chicken*! (Laughter). *Let's see whose was shorter.* (Moving to bottom of screen) *Yours was shorter three steps. Pretty good dude.*" G on her own, had maintained her focus on the issues of alternative strategies. This created some stress for her, and her laughter and informal retort was clearly a release.

This was a crucial episode, because the notion of strategy and alternative solution paths, which was unknown two weeks earlier, had become a legitimate topic of discussion. Moreover, when G posed a question (14) J felt comfortable answering. There is clearly a different consensual domain established than in the first two episodes. This episode was important to me because it marked the first time that we focused on alternative strategies.

Vignette 2. G and J had both just solved the same problem in different ways. This created an issue over whose turn it was.

Teacher: *OK. Do the very bottom one.*
G: *No, I want to do this one.*
Teacher: *We did that one all ready.*
G: *Go ahead J.*
J: *I just did it.*
G: *I did it too.*
J: *So, I could have. You're the one who took it. You were so enthused to do it.*
G: *She can.*
J: (at the same time) *You can.*
G: *You do the beginning and I'll finish.*
J: *No.*
G: *Why don't you do one is make sure you know how to do it.*
Teacher: *J you do the first step. We'll take turns.*
J: *The first step.*
G: *Yeah, we'll do step by step.*
J: *The first step is the hardest one.*
G: *All right. I'll do the first step.*
J: *No. Get out of my face.*
Both: (Laughter.)

G and J were joking about doing a problem. J's remark, "The first step is the hardest," really caught my attention. This was already implicit in G's "You do the beginning and I'll finish." I was trying to get them to focus on strategies, and here they had clearly understood enough of the structure of solving these problems to tease each other about it. G's wry sense of humor was a regular part of our discussions, especially between solving problems. This was never done at J's expense (or mine), and I found that G kept the situation relaxed and open for more constructive proposals in actual problem solving. The key feature here is that the text does not start and end with the presentation and resolution of a problem, but with the interaction of human beings.

Seeing the students twice a week for eight weeks, my goals were limited. However, I want to stress the progress we made, the difficulty, in changing a culture, and the methods that started us along. When a teacher sets up an inquiry classroom for the year, September or October classrooms must be seen as transitional cultures, and are perhaps the most serious test. It is not clear to the teacher or to the students what the rules are or what is expected. Early lessons that are designed to begin to open up the dialog in the classroom may seem a bit stilted to the observer who is expecting a full-blown inquiry classroom. Similarly in the tutorial sessions above, we are concentrating on the transformation of the culture.

Technology is usually very controlling. But technology can support inquiry. The *Workbench* provides some structure for the inquiry, and allows the teacher to turn the discussion to strategies. The computer environment is a context for a discussion. The *Workbench* has an expert system that the student can use to take a step or get a hint. There is a different feel when the "expert" suggests a goal or an action. It then becomes something the teacher and student can discuss.

X. CONCLUSIONS

Language is the means by which a student is brought into a community. As Halliday argues, "In the development of the child as a social being, language has the central role. Language is the main channel through which the patterns of living are transmitted" (Halliday [1978] p. 5). Similarly, it is the interaction of language that provides the main channel through which teachers communicate patterns of mathematical behavior to students. It is of no use if the teacher asks a mathematical question and the student can't hear it. There is a need first to develop a consensual domain, and then to learn to communicate within it. Both the student and teacher must learn to talk and listen. To be effective, the student's construction of mathematics must allow the student to carry on a mathematical discussion, that is, to participate as a member in a mathematically literate community in the classroom. The teacher's role, to create a context in which the student can construct mathematics, requires first and foremost that the teacher be a mathematically literate adult. This is distinct from the goal of having the student or teacher participating as a member of the mathematical research community. We are demanding much less. Teachers must be willing and able to participate in a mathematical discussion with students. By participating in a mathematical discussion both teachers and students approach a problem from a mathematical point of view.

NOTES

1. This research is supported by the National Science Foundation under Grant No. MDR-87 51519, "Intelligent Tools for Mathematical Inquiry." Any opinions, findings, and conclusions or recommendations expressed in this paper are those of the author and do not necessarily reflect the views of the Foundation. I would like to thank Ricky Carter, Bill Barowy, Chip Bruce, and Denis Newman for their comments on earlier versions.
2. This was very much the case in the work of the Georgia early number project (see, for example, Steffe, von Glasersfeld, Richards, and Cobb [1983]) and is characteristic of most other "constructivist" research in mathematics education [see, for example, Confrey [1988]). The constructivist effort to create models of the learner takes a snapshot, or a series of snapshots, of the learner at particular moments in time. This does not account for how the change occurs. That is, the models do not account for learning or for the activity of mathematics, which is reserved for anecdotes. The process of doing mathematics literally does not have a place in the model. Nevertheless, these models provide a framework for talking about students, and their progress. They provide teachers and researchers with a language for describing the learning of mathematics.
3. A research programme is a succession of theories which are framed by a long-term research policy which sets problems and defines areas of interest (the *positive* heuristic), and is based on a common foundation of beliefs (the *negative* heuristic) (cf. Lakatos [1970], pp. 132ff.).
4. Holding something in common is the essence of communication, as can be seen in the Latin *communicare*, to make common.
5. Converse, to engage in conversation or to talk informally, is from the Latin *conversari*, to associate with. Converse, as in reversed relation or order, is from the Latin *convertere* to turn around. It is this association with, or turning around together, that forms the basis for communication. Discuss is from the Latin *discutere* to investigate. This distinction is useful for me. I define a discussion as a conversation in which communication occurs.
6. In the Georgia project, hours were spent going over five-minute segments of children's counting behavior (cf. Steffe, von Glasersfeld, Richards, and Cobb [1983]). As our theory evolved, we came to see much richer behavior on the part of the students. It is crucial to note that all the members of the team were constantly arguing over what they were seeing. Observations are indeed theory-laden.
7. "Putting The Mind Under the Microscope." William J. Holstein and Otis Port. *Business Week*. October 10, 1988, pp. 131-2.
8. Mathematics is an instance of, in Roy Pea's terms "distributed intelligence" (cf. Pea [forthcoming]). The knowledge of the culture is incorporated in the tools of the community. A tool is a tool only in a domain in which people take it as something to be used. These tools incorporate much of the structure and knowledge of the discipline because, from the perspective of the participants, the tools are an intrinsic part of the domain. I believe it is from this perspective that we can exploit the contributions of Kaput's [1989] analysis of notational systems and his arguments that much of what is

mathematically possible is carried in the notational system.

9. See, in particular, the arguments in Beth and Piaget [1966] Chapter X.
10. This account of his discovery of quaternions is interesting because it was entered on the day of the discovery and shifts back and forth between the geometrical and algebraic problems that the quaternion resolves. This contrasts with accounts that Hamilton provided on the following day, and ten years later where the two problems are separated (cf. Richards [1988]).
11. I am deliberately ignoring formal mathematics. Although one could argue, since it is uninterpreted it is not used for communication. I would still not consider a formal language, together with the rules for interpretation as pertaining to an analysis of communication. Clearly a "formal language" of this sort can be an object of study, and can serve, as any notational system can, as a carrier of information and a tool for learning mathematics. My interest remains here with language as it is used to communicate.
12. It is inaccurate to refer to a single "mathematical research community." Researchers who specialize in different areas of mathemaatics, or even those who use different approaches within a single specialty, often have a great deal of difficulty talking to one another. This is perhaps most obvious in the dynamics of delivering a paper. Usually, the first ten minutes are spent identifying the problem and underscoring its significance. This is intended for the general math research community. The speaker then coughs twice and dives into the details of the problem that is directed at those who have experience with the particular specialty. For the sake of this paper, I refer to a single math research community, perhaps the one to which the first ten minutes of the talk is directed.
13. In Richards [1988] I describe several episodes in the history of mathematics where these sorts of cognitive shifts occurred.
14. Cf. Stolzenberg [1978] for a careful analysis of this act of acceptance.
15. Lemke [1982] carefully analyzes a variety of these situations in the context of science classrooms.
16. In fact, the real situation may be even worse than this characterization. In the classic research on the dynamics of questioning in science classes by Mary Budd Rowe [1974], she reports that the time allowed after a teacher's questions for a student's answers was less than 0.9 seconds.
17. The caricature of this process in Lakatos [1976] provides insight into the need for argument in science and math. The process of argument is not a flourish, it is an essential aspect of the individual's construction of mathematics. As Bloor argues, "Negotiations create meanings." ([1976], p. 130)
18. The importance of this insight is often overlooked. In a survey of 87 year-seven math classes, observed in 37 schools, Barbara Clarke et al. [1989] report the following distribution of time given to various methods of instruction: blackboard 42% (teacher 25%, students 17%), text 37%, worksheets 14%, activities (games/manipulatives) 4%, and discussion 3%.
19. The direct sharing of the teacher's role is also demonstrated in reciprocal teaching for reading instruction (Palinscar and Brown [1984]) where students and teacher alternate roles by taking turns as teacher. Whoever assumes the teacher's role formulates questions, constructs summaries, and makes a prediction or clarification. Again, from my perspective, the success is directly due to altering the context of the discussion in the classroom.
20. In our judgment, intelligent computer tools like the *Algebra Workbench* can assist the teacher in creating such an environment. These provide more structure than nondirective computer environments, called *microworlds*, and depend intimately on the teachers involvement. Microworlds, in contrast, have been created to enable students to probe

and study the behavior of mathematical or physical systems and to discover and extend their knowledge through such exploration. Microworlds can be used productively in this way by sophisticated students who already know how to learn.

21. Programming languages with recursive functional constructs are the perfect instrument for supporting a constructivist approach to developing mathematics. As is the case with microworlds, however, even the most accessible of such languages, Logo, requires a supporting structure to enable students to learn to use it easily and effectively.
22. Taken from Prince [1970], pp. 58 *ff*.
23. This name was suggested by Goldenberg and Feurzeig [1988].

REFERENCES

Albers, D.J. and G.L. Alexanderson (1985). *Mathematical People: Profiles and Interviews.* Chicago: Contemporary Books.

Beth, Evert W. and Jean Piaget (1966). *Mathematical Epistemology and Psychology.* Dordrecht, Holland: D. Reidel.

Bloor, David (1976). *Knowledge and Social Imagery.* London: Routledge and Kegan Paul.

Brown, G. Spencer (1969). *Laws of Form.* London: Allen and Unwin.

Brown, John S. and Allan Collins (in press). "Cognitive Apprenticeship: Teaching Students the Craft of Reading, Writing, and Mathematics." In L.B. Resnick (ed.) *Cognition and Instruction Issues and Agendas*, Hillsdale, NJ: Lawrence Erlbaum Associates.

Carlsen, William S. (1989). "Questioning in Classrooms: A Sociolinguistic Perspective." Paper presented at American Educational Research Association, San Francisco, CA: March 27–31, 1989.

Clarke, Barbara, Doug Clarke, Charles Lovitt, and Max Stephens (1989). "Expanding the Repertoire of Pre-Service Mathematics Teachers." Paper presented at the National Council for Teachers of Mathematics, Orlando, FL, April, 1989.

Confrey, Jere (1988). "The Concept of Exponential Functions: AStudent's Perspective." Athens, GA.

Erickson, Frederick (1980). "Timing and Context in Everyday Discourse: Implications for the Study of Referential and Social Meaning." *Sociolinguistic Working Paper, No. 67*, Austin, TX: Southwest Educational Development Laboratory.

Feurzeig, Wallace, John Richards, and Nancy Roberts (1987). *Intelligent Tools For Mathematical Inquiry.* Proposal submitted to Applications of Advanced Technology, NSF.

Gleick, James (1987). *Chaos: Making a New Science.* NY: Viking.

Goldenberg, E. Paul and Wallace Feurzeig (1988). *Exploring Language with Logo.* Cambridge, MA: MIT Press.

Gumperz, John J. (1981). "Conversational Inference and Classroom Learning." In Judith Green and Cynthia Wallat (eds.), *Ethnography and Language in Educational Settings.* NJ: Ablex, 3–24.

Hadamard, Jacques (1949). *The Psychology of Invention in the Mathematical Field.* Princeton: Princeton University Press.

Halliday, M.A.K. (1978). *Language and Social Semiotic.* Baltim. MD: University Park Press.

Halliday, M.A.K. and Ruqaiya Hasan (1976). *Cohesion In English.* London: Longman.

Hamilton, Sir William Rowan (1843). Notebook 24.5 entry of 16 October 1843. In *Mathematical Papers Vol. III*, Algebra. H. Halberstam and R.E. Ingram (eds.) Cambridge: Cambridge University Press, 1967: 103–5.

Kaput, James (1991). This volume.

Kuhn, Thomas (1963). "The Function of Dogma in Scientific Research." In A. Crombie

(ed.), *Scientific Change*, NY: Basic Books.
— (1970). *The Structure of Scientific Revolutions*, 2nd edition. Chicago: University of Chicago Press.
Lakatos, Imre (1976). *Proofs and Refutations*. John Worrall and Elie Zahar (eds.) Cambridge: Cambridge University Press.
— (1970). "Falsification and the Methodology of Scientific Research Programmes." In I. Lakatos and A.E. Musgrave (eds.) *Criticism and the Growth of Knowledge*, Cambridge: Cambridge University Press, 91–106.
Lampert, Magdalene (1986). "Knowing, Doing, and Teaching Multiplication." *Cognition and Instruction*, 34, 305–42.
— (1987). "Choosing and Using Mathematical Tools in Classroom Discourse." In J. Brophy *Teaching for Meaning Understanding and Self-Regulated Learning*, JAI Press.
Lemke, J.L. (1982). "Talking Physics." *Physics Education*, Vol. 17, 263–7.
Lochhead, Jack (1985). "Teaching Analytic Reasoning Skills Through Pair Problem Solving." In Judith W. Segal, Susan F. Chipman and Robert Glaser (eds.), *Thinking and Learning Skills*, Volume 1, Hillsdale, NJ: Lawrence Erlbaum Associates, 109–31.
Maturana, Humberto (1978). "Biology of Language: The Epistemology of Reality." In G.A. Miller, and E. Lenneberg (eds.), *Psychology and Biology of Language and Thought*, NY: Academic Press, 27–64.
Mehan, Hugh (1979). "The Competent Student." *Sociolinguistic Working Paper*, 61. Austin, TX: Southwest Educational Development Laboratory.
Oakeshott, Michael (1962). "The Voice of Poetry in the Conversation of Mankind." *Rationalism in Politics*. NY: Basic Books.
Palincsar, Annemarie S. and Ann L. Brown (1984). "Reciprocal Teaching of Comprehension Fostering and Monitoring Activities." *Cognition and Instruction*, 1, 117–25.
Poincaré, Henri. *Science and Method* (1908). Trans. G.B. Halstead, *The Foundations of Science* (PA: Science Press, 1913), 359–546.
Polya, G. (1971). *How To Solve It*, Second Edition. NJ: Princeton University Press.
Popper, Karl (1959). *Logic of Scientific Discovery*. English translation of *Logik der Forschung*, 1934. London: Hutchinson.
Postman, Neal and Charles Weingartner (1969). *Teaching as a Subversive Activity*. NY: Dell Publishing.
Prince, George M. (1970). *The Practice of Creativity*. NY: Collier Books.
Richards, John (1988). "Epistemology and Mathematical Proof." *Methodologia* 3, 73–121.
Richards, John and Wallace Feurzeig (1988). "Intelligent Tools For Algebra." *Technology and Learning*, 2: 1–11.
Rowe, Mary Budd (1974). "Wait-Time and Rewards as Instructional Variables, Their Influence on Language, Logic, and Fate Control: Part 1. Wait Time. *Journal of Research in Science Teaching*, 11, 81–94.
Schoenfeld, Alan H. (1983a). "Problem Solving in the MathematicsCurriculum: A Report, Recommendations and an Annotated Bibliography." *The Mathematical Association Of America*, MAA Notes, No. 1.
— (1983b). "Episodes and Executive Decisions in Mathematical Problem Solving." In R. Lesh and M. Landau (Eds.) *Acquisition of Mathematical Concepts and Processes*, New York: Academic Press.
— (forthcoming). "On Mathematics as Sense-Making: An Informal Attack on the Unfortunate Divorce of Formal and Informal Mathematics." In D.H. Perkins, J. Segal, and J. Voss (eds.), *Informal Reasoning and Education*, Hillsdale, NJ: Erlbaum.
Schwartz, Judah, Michal Yerushalmy, and Myles Gordon (1985). *The Geometric Supposer Manual*. NY: Sunburst Communications.
Steffe, Leslie, Ernst von Glasersfeld, John Richards, and Paul Cobbs (1983). *Children's Counting Types*. NY: Praeger Publishers.

Stolzenberg, Gabriel (1978). "Can an Inquiry Into the Foundations of Mathematics Tell Us Anything Interesting About Mind?" In G.A. Miller and E. Lenneberg (eds.), *Psychology and Biology of Language and Thought*, NY: Academic Press, 221–69.

Whimbey, A. and Jack Lochhead, (1980). *Problem Solving and Comprehension: A Short Course in Analytical Reasoning* (2nd ed.). Philadelphia: Franklin Institute Press.

Winograd, Terry and Fernando Flores (1986). *Understanding Computers and Cognition.* NJ: Ablex.

NOTATIONS AND REPRESENTATIONS AS MEDIATORS OF CONSTRUCTIVE PROCESSES

JAMES J. KAPUT

I. INTRODUCTION

* Why, despite the many sources of difficulty, is successful communication of mathematics among individuals possible?
* Why is it that to apply mathematical ideas, one must inevitably choose one or more notations in which to materialize those ideas?
* And why the large variation in the ways that these notations support and/or constrain our thinking processes?

In this chapter I will not presume to answer such questions, but rather will attempt to provide some means for others, hopefully including the reader, to gain insight into them. Central to this task will be to describe the twin mediating roles of notation systems, in mediating between what is normally regarded as "pure mathematics" and one's experienced world, and in mediating communication processes among individuals. In so doing, I hope also to show how a representational framework for mathematical cognition and learning is consistent with constructivism.

The latter objective is actually primary, because a representational perspective is sometimes seen as inconsistent with constructivism in two ways. First, the understandable tendency, based on a long epistemological and semiotic tradition, is to think that if "X represents Y" in any of several senses of "represent," some of which will be teased apart later, then somehow Y exists as a referent in some absolute sense – an assumption antithetical to constructivism, and one which I do not make. Instead, I regard "representations" as "re-presentations" of one aspect of one's experience by another. Second, constructivism is often seen as underplaying the role that language and other symbolic forms inherited with our cultural and linguistic traditions play in mental activity, especially in social contexts. I leave social interactions, especially those involving natural language, to others in this volume, and instead concentrate on the key role that notations play in mathematical constructive processes.

The paper is in two parts, the first sets a theoretical frame and the second applies it to a concrete example of notations used in the solution of a common mathematical problem involving the sums of integers. The overall intellectual strategy in development of this framework is to pull apart, for purposes of

E. von Glasersfeld (ed.), Radical Constructivism in Mathematics Education, 53–74.

analysis, different aspects of the learning and doing mathematics that are usually experienced as an integrated whole. In particular, I shall separate (mental) conceptions from (physical) notations, acts of reading notations from acts of writing notations, and consensual worlds of "givens" from subjective and material worlds. In Part II I attempt to show how they hold together.

PART I: A THEORETICAL FRAMEWORK

The Preliminary Framework

Perspectives: Who is Looking at What?

I take the perspective that while a person has no direct knowledge of any surrounding environment, an observer of that person can view and comment on (using the observer's description of) that person interacting with (the observer's version of) that person's *experienced* environment. Figure 1, adapted from (Kaput, 1985), depicts the view that I am taking. I am especially interested in situations where the observed person is interacting with mathematically oriented notations of some kind. I regard these notations as being part of the user's world as experienced and as *contributing to the organization of that person's thinking processes* – but not as (perhaps imperfect) mental representa-

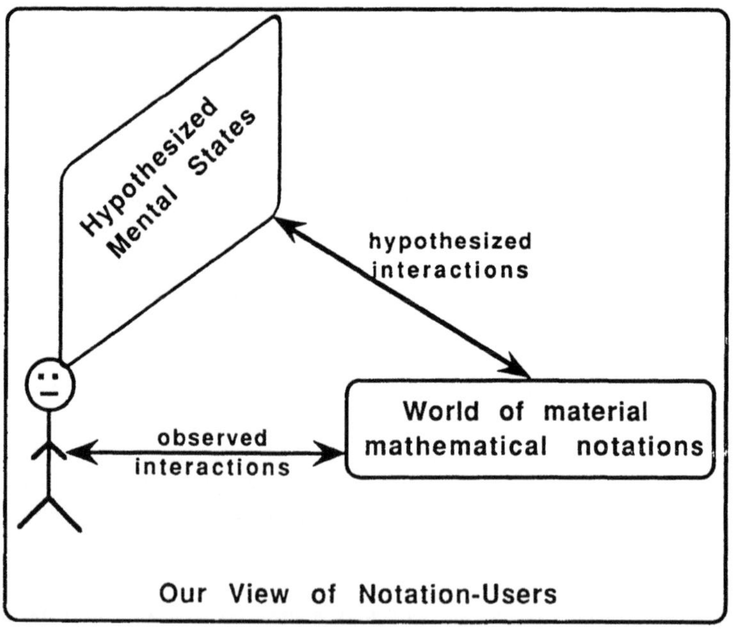

Figure 1.

tions of an external world. Of course those material notations do not instantly or automatically organize a person's experience – their organizing power is the result of constructive processes using the records of prior experience. Quite obviously, a person cannot use a coordinate graph to describe or reason about some phenomenon until he/she learns how coordinate graphs work, which depends on knowing something about how number lines represent numerical data, which in turn depends on knowing something about number, order, and so on. But once cognitively "available," *as generative mental operations in the user*, particular notation systems play a critical role in determining the structure of a person's thinking processes. But how do the particular features of a notation system function cognitively – how do material notations and (our hypothesized) mental constructions of the user interact to produce new constructions? Mathematics educators need answers to such questions in order to fashion learning environments and situations that exploit those environments.

1.2 Notation Systems as Architecture

One can draw an analogy between the way the architecture of a building organizes our experience, especially our physical experience, and the way the architecture of our mathematical notation system organizes our mathematical experience. As the physical architecture constrains and supports our actions in ways that we are often unaware of, so do mathematical notation systems. Furthermore, until recent times much architectural design was the product of evolutionary functionalism rather than deliberate design. In modern societies the design considerations underlying architectural functionality are now made and applied explicitly by specialized architects. Our role as analyzers and designers of mathematical representation systems, especially in computer contexts where we have greater control of what the user can experience, is, or should be, increasingly analogous to that of architects (see Hooper, 1986).

One can extend the analogy to the epistemological level as well. After all, notation systems are virtually always physically instantiated, and it is there, in the material realm, that they influence us as subjective beings. In Part II I shall examine one case in some detail to illustrate the kind of analysis that I think can be fruitful in this respect. But before doing so, I must clear some theoretical ground.

1.3 How to Describe Representational Acts?

I have distinguished between:

(1) *mental structures* as means by which an individual organizes and manages the flow of experience, and
(2) *notation systems* as materially realizable cultural or linguistic artifacts shared by a cultural or language community.

When *materially instantiated*, as marks on paper, sounds, or electrically based signs in a computer display, notation systems are used by individuals to organize the creation and elaboration of their own mental structures. They do this by structuring physical records of prior mental activity and by structuring actions on those records, both physical and mental actions.

I wish to draw another analogy, between the idea of a notation system in mathematics as a specifiable organizing structure and the idea of grammatical structure in natural language. Morrow (1986), expressing a traditional line of thought in linguistics, notes the following regarding the development and roles of grammar structures as they occur in natural language: "Distinctions that help organize knowledge because they occur so frequently and widely across many different conceptual domains are coded by grammatical categories, which participate in the most regular, obligatory part of language." (p. 425) Morrow cites Sapir "who distinguished content words from relational parts of a sentence .... He argued that the content words express the 'material content' of sentence meaning, whereas the latter express the 'structural mold'..." (p. 426). He further suggests that the relatively small number, great generality, and regularity of grammatical constructs, and not the huge variety of content words, provide the means by which limited cognitive resources can be efficiently applied to organize and communicate the enormous richness and continually changing flux of experience.

Similar considerations apply to the notational systems used across mathematics to express relationships of infinite variety and enormous generality. While relatively few in number, they combine with one another and with non-mathematical representational forms to support a startlingly rich array of mathematical utterances, by individuals and across history (Cajori, 1929,a,b). However, the very slippery form/content distinction becomes even more so in mathematics, where the growth of mathematics (both ontogenetic and phylogenetic) is bound up with repeated transformations of form at one level becoming content at the next (see, Kaput, 1987b, for a series of examples).

1.4 The Signifier-Signified, Notation-Conception Referential Relation

As a first approximation I depict the relation between mental representations and physically instantiated representations as in Figure 2. In an obvious assumption of referential relationship, the lower level of this diagram is often referred to as the "signifier" and the upper level as the "signified" (Vergnaud, 1987). Signifiers, that is, material notations, include physical apparatus as well as physically instantiated symbols – not only alpha-numeric, but pictorial, diagrammatic, and even aural notations when the medium is sound. Such notations can be consensual, as with natural language or coordinate graphs, or they can be idiosyncratic, personal marks. In either case, they are the things that we read (interpret) and write (draw, say, etc.)

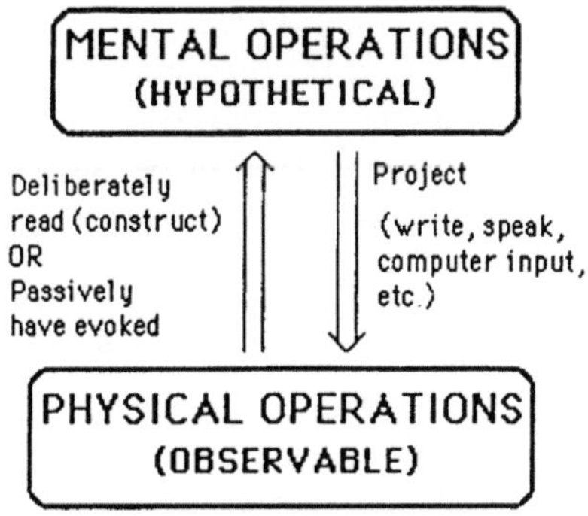

Figure 2.

When using such material notations, we build and/or elaborate our mental structures in cyclical processes that go in opposite directions. The arrow pointing upward is intended to depict two types of processes: deliberate reading, and the more passive, less consciously controlled and less serially organized processes of having mental phenomena *evoked* by the physical symbols. (The former take place on the long side of the 100 millisecond boundary while the latter take place on the short side (Hofstadter, 1980).) The arrow pointing downward also depicts two types of processes: the act of projecting mental contents either onto existing symbols, and the act of producing new ones, which includes the physical elaboration of existing ones. (A similar division into short and long processes seems to apply here.)

Although projections occur in the reading and evoking as part of the underlying cyclical processes matching percepts and concepts (Grossberg, 1980), I distinguish downward oriented ones from upward oriented ones on the basis of where the process gets its major impetus. In the downward oriented case, one has some conceptual structure or operation that one seeks to externalize for purposes of communication or testing. The latter might occur when one has some vague or incomplete visual image and wishes to clarify it by comparing it with some existing drawing or producing a new one. Upward oriented processes are based on an intent to use the physical material to assist one's conceptual activity in traditional acts of reading and interpreting (which inevitably include passive acts of evocation mentioned above).

Note that projections need not amount to writing in the usual sense, but can include acting on virtually any kind of physical system or apparatus, where the system itself determines the kind of "writing" that it can accept. For example, writing might take the form of placement of cardboard puzzle pieces, turning a crank on some physical apparatus (e.g., Greeno, 1988), manipulating Dienes Blocks, drawing on a computer screen, etc. Below we shall extend this layout to help describe representation acts involving more than one system of notation.

1.5 Consensual Domains: The Third World

We need to set some additional epistemological assumptions. Figure 3, twisting an idea from Sir Karl Popper (1972), poses the existence of three "worlds," (1) the world of "material stuff," (2) the world of mental events (Popper refers to it as the "subjective world"), and (3) the world of "consensual stuff." Popper treats World 3 as the "world of objective knowledge" (pp. 73–4) in his attempt to rescussitate objectivity from the body blow inflicted by Hume. Here I am treating it as the consensual domain, i.e., the variably bounded repository of what a particular cultural and language community takes as true on a particular occasion. That community can be of virtually any size larger than one. For a large intellectual community such as an academic discipline at a conference, it includes such things as the currently accepted body of knowledge, perspectives and methods constituting that discipline. Inquiry and dispute take place at the borders, which change over time.

In the case of a mathematics classroom, growth and change in the consensual domain is (or should be) rapid and at the heart of what is important. For a small group working on a common problem, such temporal variation is even more rapid. In fact, the stability of the consensual world depends very much on the extent and nature of the group.

If the consensual community is that of mathematically competent adults, then this consensual world includes the usual mathematical "concepts," e.g., linear function and ratio. It also includes the usual consensually applied mathematical notation systems, our means of designating such "concepts," e.g., algebra, coordinate systems, number systems, and so on.

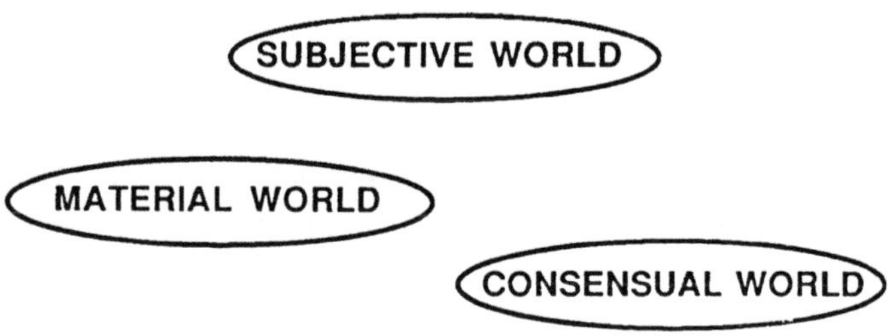

Figure 3.

By relativizing our characterizations of knowledge to cultural and linguistic communities, I render it an unattractive and inadequate foundation for all knowledge, especially among professional philosophers. But I am not looking for the foundational turtle on which all knowledge can comfortably sit. Rather, I am searching for means to help describe, as systematically as possible, how people use *some* aspects of their worlds (the notational ones) to represent *other* aspects of their worlds to themselves for purposes of thinking, and to others for purposes of communication. The consensual world in this framework is intended itself to be a notational device. The flexibility of this device lies in the variation accorded the definition of "cultural and linguistic community," which depends on the community we happen to be dealing with in our descriptions of what I call "representational acts."

Beware of a likely tendency to interpret the material world as the place where "reality" is to be found, the ontological bedrock. This is not what I intend. Rather, I intend that it stand for what is experienced as physical, just as the top of Figure 3 stands for what is experienced as mental – or describable by others as subjective experience. In particular, the physical world is not assumed as a fixed, universally available entity that our subjective experience attempts, with varying success, to represent or model. (I am not advocating naive realism here.) Instead, the subjective and the physical worlds are best thought of as different domains of experience. The consensual world helps us discuss those aspects of experience, either material or subjective, that are the product of intersubjective processes, aspects that are in the realm of negotiated meanings. Such include both short term meanings as occur in ordinary conversation as well as the results of long term historical and cultural processes – what Popper (1972) regards as the world of "objective knowledge." Of course, a key feature of our larger consensual world that includes our language and customs, is its role in defining itself. Enculturation includes not only the building of a world, but the building of means for building and elaborating that world (Goodman, 1984).

1.6 A Second Dimension of Reference

Figure 4 elaborates the framework further, filling it out to include a horizontal dimension of reference – which, as a species of representation, is quite different from the vertical signifier-signified one. Here, I intend to express relationships between "notation A and referent B" where each (and perhaps even the correspondence) is expressible in material form – but where *the actual referential relationship exists only in terms of the mental operations of members of a particular consensual domain.* The reference between A and B exists by way of composite actions that "pass through" the subjective world – as combined acts of interpretation, mental operation, and projection to a physical display. The items A and B depict physical observables – for example, an equation A and its graph B both on a piece of paper or a computer screen, or a numeral A that refers to the numerosity of a set of objects B. The mental operations might include counting the set of objects, for example. However, we must remember that the items at the cognitive level depict hypotheticals, exactly as in Easley's "wavy line diagrams" that Clement, for example, has used so deftly to describe clinical data (Clement, 1982). The directionality of the reference depends on the cognitive operations involved, which in turn depend on the context, and hence is not fixed. At one time B may act as a referent for A (e.g., as above, the numeral A acting to represent the numerosity of a given set B), and at another the situation may be reversed. The numeral B may be given in advance and one counts out a set of objects A to represent the numeral.

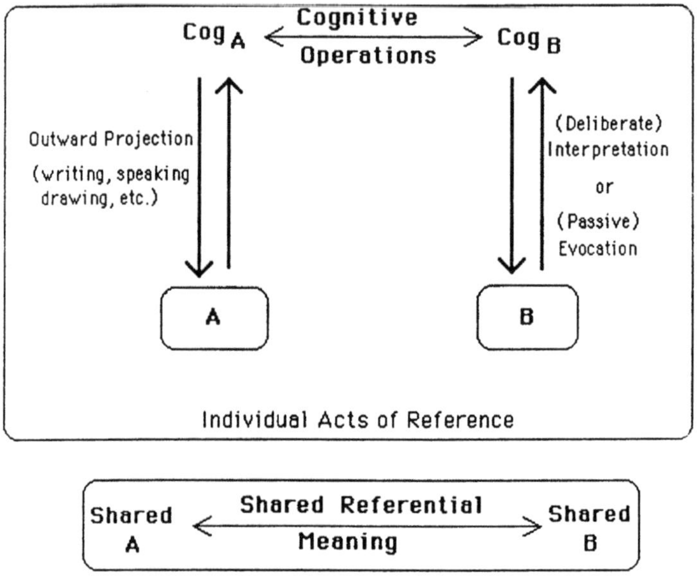

Figure 4.

The bottom-most level of the diagram is intended to be used to describe a representation situation for a particular community, as when one says "algebraically, the quadratic function $f(x) = x^2 + 1$ [A] represents the parabola [B]" to a community of appropriately knowledgeable mathematics teachers; or when one describes a piece of physical apparatus as representing a linear function $y = aX + b$ for the same community (Greeno, 1988). By focusing on the physical and intersubjective, one suppresses cognitive aspects of referential relationships. But if one wishes to describe particular representational acts by individuals, one *must* take into account the cognitive level. The bottom-level serves both as a reminder that certain aspects of the situation may be intersubjectively shared and as a means for being explicit about exactly who is in the sharing community. Some of my earlier discussions of the representational nature of mathematics (Kaput, 1987a) can be regarded as a description involving the lower two thirds of the diagram – the physical and the consensual.

The notion that the actual functioning of material notations can be discussed only in terms of interpretants is in keeping with a more general twentieth century trend towards interactionalist/transactionalist epistemologies (Toulmin, 1972, 1982). While this is not the place for an historical recounting of this cultural phenomenon, I should note that C.S. Peirce (1933) described semiotic behavior as involving an interaction among "sign, object, and interpretant" (in our terms, A, B, and the mind in which the integration of Cog_A and Cog_B takes place, respectively). In the context of the development of a symbol's referential meaning, Werner and Kaplan (1962) phrase a similar view, as the development of the "linking of a symbol and referent with the same organismic state" (p. 18). Von Glasersfeld (1987) puts it this way: "A representation does not represent by itself – it needs interpreting and, to be interpreted, it needs an interpreter" (p. 216).

To be interpretable, and hence be available to function in supporting of reasoning, features of notations A must be perceivable and relatable to other prior knowledge. Under these circumstances, I shall refer to a feature, or even a notation system, as being *available* for a particular interpretant. Of course, to be available for interpretation does not guarantee that the interpretation will match the consensual one, but only that it will be "seen," i.e., that it results in the activation of mental activity.

1.7 Notation Systems as Abstractions and the Matter of Platonism

Another factor relating to the character of notational power has to do with degree of abstractness. Notational systems such as the two-dimensional coordinate graph system, two column tables of data, and the system of two variable algebraic equations exists as abstract entities in both the shared mathematical culture and the mental structure of competent mathematicians. However, I am using the word "abstract" in a non-platonic sense. In Kaput (1989) I distinguish between a "notation system" and an "application" of that system. An application of the coordinate graph system, for example, is a particular graph, a particular set of points, say an ellipse, in a physical medium. The coordinate graph system, on the other hand, can be regarded as existing apart from any particular application of it, either as a cognitive or a culturally shared entity. For those in the consensual community who share it, it is a particular way of organizing information, with its own rules of syntax and so on. Furthermore, it can even be physically instantiated apart from a particular application, e.g., as a pair of coordinate axes on a piece of graph paper.

I draw a strong distinction between this point of view and old fashioned mathematical platonism, whereby mathematical objects are presumed to exist in some abstract sense apart from any presentations of them. I agree with bishop Berkeley in the position that we can know nothing apart from some physically instantiable features to identify it. This includes knowledge of entities with no concrete physical reality, e.g., dancing pink unicorns, or the idea of hubris. For example, one cannot know the concept of function (or number) independently of all the different ways we have of representing and using functions (or numbers). What is sometimes thought of as the "abstract concept of function (or number)" is better regarded as a notationally rich web of representations and applications of functions (or numbers) in the classical sense of "application" as a model of some situation. Psychologically, this amounts to having a rich set of mechanisms available in long term memory to generate and apply functions (or numbers), which simultaneously means that one can call any of these to mind as needed and write them in an appropriate physical medium. Another psychological aspect of such abstract understanding is that it requires relatively few concrete features to be generated cognitively and used. Thus to know a mathematical idea "abstractly" means to have a sufficiently rich set of mental structures so as to be able to deal with the idea on the basis of relatively few salient features either in a notation or in a situation to be modeled. This contrasts with the typical neophyte's "concrete" understanding, which is context-bound and embodied in a single notational system. Either case, however, requires *some* material forms and embedded features for it to be generated.

PART II: APPLICATIONS OF THE FRAMEWORK

A Particular Example – Deriving a Formula for Sums of Integers

Let us consider a particular extended example of some referential acts that are typical of what occurs in classrooms and problem solving that uses multiple modes of representation. Indeed, versions of the following have occurred in a class for academically disadvantaged freshmen that I have taught for more than ten years, as well as in other classes. The task at hand, one familiar to teachers of mathematics, is to come up with a general formula for the sum of the first n consecutive integers.

2.1 Inferences from a Table of Data

One way is inductive: set up a table of data with consecutive integers running down the left column and, in the corresponding position in the right column, the sum of integers up to that point. Then use whatever means one has available to describe the data in the right column as a closed form function of that in the left column. For our purposes here, a distinctive characteristic of this approach is its reliance on the repeated elaboration of a single formal notation system B – the numerical table of data – to produce a formula in another formal notation system A – algebra. At a gross level of description, we build and reason about B in Cog_B until we are able to build a Cog_A that we then write in material form as a formula. Then we can regard A as representing B – in fact, the generality of the formula A covers all extensions of the table, not merely those that supported our cognitive operations leading to Cog_A, hence A. Of course, in practice this process is likely to require several cycles.

2.2 Reasoning with Concrete Materials

A pedagogically and illustratively richer example involves the use of concrete materials, say "staircases" of Cuisenaire rods B, to build A. This example will also help illustrate how the particulars of the physical system help support and constrain the cognitions needed to build the required formal representation, and how the particulars of one notation match and fail to match those of another notation.

First of all, the problem is stated initially as a sum of integers problem, often embellished with the tale of young Gauss being told to find the sum of the first 100 integers and coming up with the answer in much less time than anyone, and especially my students, expect. In the dozens of occasions when I have used this approach over the years, in both didactic teacher-centered as well as exploratory laboratory classes, the following issues are dealt with. I shall use Figure 5, an expansion of Figure 4, to help with the discussion.

First of all, the problem has to do with objects, namely positive integers, that I assume are in the consensual domain for the college freshmen at hand. (The consensual domain is not depicted in Figure 5.) These objects are physically instantiated as numerals in the usual notation system, the "C" in Figure 5. The numerals are written in some orderly fashion, say in a table of data. I regard integers cognitively as abstract (meaning that they are stripped of perceptual features) quantities phenomenally experienced as abstract objects, variously abstracted/interiorized from actions on physical quantities, with properties such as order, replicability, and additivity variously abstracted from actions on the physical ones in the usual Piagetian sense (Steffe, this volume). Thus the cognitive correlate Cog_C associated with the numerals and the integers are conceptions of quantities and relations among them. The assumption that the integers are shared means that we can assume reliable communication, but not that we know exactly what the conceptions are, and, in fact, conceptions may differ significantly from student to student.

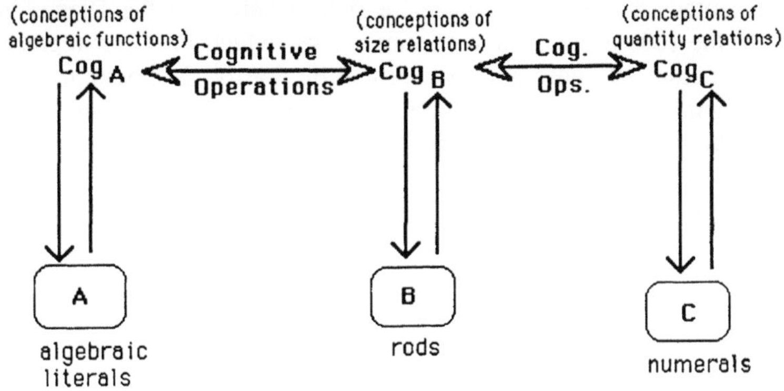

Figure 5.

The rods [B] are introduced (if they are not already familiar) as corresponding to the positive integers. To be more accurate, the referential relationship, the correspondence, is between rod *lengths* and (positive) integral *quantities.*

Indeed, the relation is typically introduced with something like the following: This rod stands for one; this red rod, twice as long, stands for two; this light green one is three. All this is done complete with gestures, holding up the rods together, etc. (Note that initially the integers are presented (materialized) orally, in the sound medium, and only later, in the course of the reasoning process, do they appear in the paper-pencil medium.) The length-integer correspondence is not quite explicit, and very early into the introduction a rod-integer correspondence comes to stand in its place (when the green rod is identified with three). In fact, the length feature quickly comes to be the only relevant feature, so that color, volume, mass, etc. are all ignored.

Now the cognitive/perceptual process of selecting features and coordinating these with features of previously existing cognitive entities takes place, of course, at the mental level of the diagram. It proceeds quickly and easily, as Gattegno and others recognized many years ago, because the students have a ready and stable cognitive structure built from experience with sticks and straight lines. (In fact Gattegno, Cuisenaire and others have shown, conversely, that one could start with Cog_C at an early age and build Cog_B out of Cog_C.) Thus the integration of students' conceptions of integers Cog_C with that of rods Cog_B is unproblematic, and we can say that "rods represent integers" regarded as conceptual entities via a vertical route from B to Cog_B (a "reading" act whose major component is feature selection) over to Cog_C (the integration act). If we then write an integer as a numerical C, we can regard the corresponding rod as representing the written numerical via the triple composite B to Cog_B, over to Cog_C and then down to C. It is this triple composite that constitutes the referential relationship between numerals and rods. Incidentally, using the word "composite" in a somewhat different way, I should note that the composite nature of the integers (as composite units in the sense of Steffe, et al, 1988), is not represented in the usual Cuisenaire rods, whereas in the Dienes version ("longs"), the rods are etched with unit marks, which thus can represent the composite nature of integers more explicitly. Lastly, I should mention the vertical signifier/signified referential relationships, for example, between abstract quantities (conceptual entities) and written numerals.

2.3 Actions on Notations and Their Records: Incrementing Integers and Rods

Of course, the purpose of building a correspondence between items is to extend it to correspondences between *actions* on the items. Here is where payoffs in reasoning are to be had. The first actions on items in B and C, respectively, are the two notational forms of incrementing by one. Each of these in turn has two forms. In one of these the *historical record* of the incrementing is preserved, whereby the earlier numerals remain in position as a sequence in a table of data, or the earlier rods remain in position, most typically as a "staircase" – as illustrated in the lower right side of Figure 6. Here, in B, the linear order

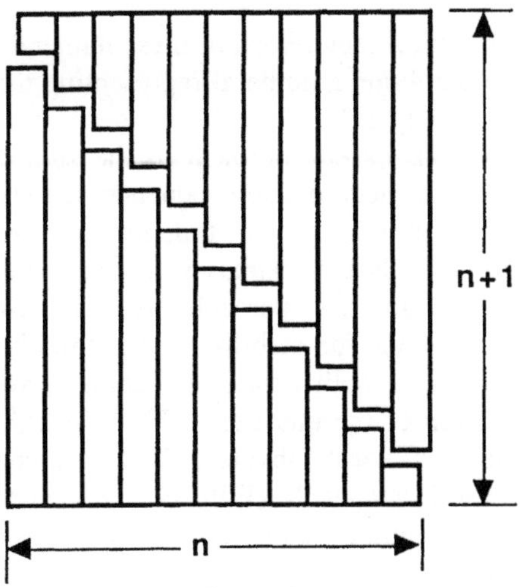

Figure 6.

relation among the items can be imposed on the elements of the linearly growing set of rods, because the rods can explicitly represent the quantitative content of integers, that is, their relative size. In this case, of course, we must assume it exists in some form in Cog_B – presumably inherited from Cog_C, conceptions of positive integers, where quantity and order are "stored," probably as general processes.

In C the record takes the form of a sequence of numerals whose only order is that inherited from the order in which they were produced. Their relative size as abstract quantities can be keyed to this order, but is not explicitly marked in the numerals themselves. The rods mark order and size in a way that the numerals do not. For example, the numeral "6" is not larger than "5" whereas the rod corresponding to the (shared and cognitively experienced) integer 6 is longer than the rod corresponding to the integer 5. I should remark that the particular physical medium in which a conceptual item is written plays an important role in determining whether an historical record is preserved. If it is sound, then the record is not preserved, but dissipates almost instantly.

The other form of incrementing overwrites or discards the previous item, and so only the latest rod or numeral is displayed in material form. This type of incrementing does not yield or represent the new items that we need to examine, namely the sums of integers, so will not be pursued further.

Actions on notations produce state-changes in those notations. As we have just seen, the state changes may be accumulative or not. That is, old states may or may not be included in the new states as they are produced, respectively. In the accumulative version of the incrementing action, in both rods and numerals we produce the basis for further progress on our goal, determining a general formula for the sum of consecutive integers. In C it can be a sequence of numerals arrayed in a table of the type previously discussed, e.g., as items in the left hand column of a vertical table. Each of these numerals can have associated with it in the right hand column a numeral representing the sum of the integers up to and including that integer.

At the heart of the rods-based approach to the sum of integers problem, however, is the fact that the accumulated sequence of rods arrayed in the form of a staircase explicitly provides, via students' well established cognitive structures associated with rods and staircases, an experiential item on which to base the sought-after sum. Its perceptually and cognitively available "objectness" seems important for stabilizing our thinking about sums. It is usually straightforward to establish the connection between sums and rods, usually with some version of the question, "What does the total length of the rods in a staircase stand for?" or its obverse, "How can we use the rods to stand for the sum of the first 6 integers?" (With college-age students it has never seemed necessary to disassemble the staircase into a "train" and represent the sum more explicitly as a concatenation of rods, although I assume that younger children might not be able to connect their conception of sum with the staircase as easily and thus might require some intermediate steps of this sort.) We are thus able to make an important jump, and ask the question "What does this double staircase stand for?" (See Figure 6.) The answer "double the sum," is important, because we are now able to compute the double sum using the two staircases combined as in Figure 6 very easily. Here, of course, we are again acting on B, replicating it and structuring the new state in an especially fruitful way. Furthermore, with respect to Figure 5, the action on B is made to correspond with an action in Cog_C, forming the sum.

One can compute the total length of the rods in the double staircase, now a rectangle, in either of two ways: either by asking how many vertical rods (actually, rod-pairs) of a particular length appear and thus what is their total length (the product of the number of rods and their length); or by using the fact that the double staircase is a rectangle whose area corresponds to the total length of its constituent rods. The leap to either of these conceptualizations – especially the latter – sometimes requires some prodding, usually in the form of leading questions. Note that the area interpretation is an application of selecting only the length of the rods as their representing feature. Again, to perform the appropriate computation requires selecting an available feature in the new B, in this latter case the rectangularity of the resulting configuration, applying prior knowledge (brought to bear in Cog_B) concerning areas of rectangles, and applying that knowledge to the creatively organized (by the teacher) and perceived (by the students) set of rods (in our shared, consensual domain).

2.4 *Particularity and Generality*

To this point, and especially in the way I cheated by using "n" in the labels of Figure 6, I have ignored the large differences in capacity for embodying generality of our three notations (A, B, and C). This issue is quite complex, and

I will certainly not exhaust its content here. There appear to me to be three conceptual factors, and a number of notational factors to be considered in our situation. And each of these factors may relate differently to acts of interpretation (reading) versus acts of externalization (writing).

First, conceptually, generality seems to depend on ability to recognize instances as members of a class, an ability that is usually assumed to be subject to development and maturation. Second, the conceptual status of the class itself may be problematic for individuals. If the class is an infinite set, the set of all integers for example, then its conceptual existence may be questionable for certain individuals, including college students (Fischbein, Tirosch, Hess, 1979; Wheeler & Martin, 1988). Furthermore, the conventions for expressing this infinitude may vary considerably depending on the notation system being employed. In A it is integrally built into the structure of the system itself. In C (tables of data), specific conventions support general statements, e.g., ellipses (three dots) at the end of a table. In the rods B, there are few conventions, so that I normally must accompany the physical rods with certain orally provided verbal amplifications, and perhaps even "cheat" as I did with the labels in Figure 6, where I overlaid portions of the notation system A onto B. However, we should remind ourselves that Figure 6 is a *picture* of rods, and is not composed of rods itself. By nature, physical objects seem to pull towards the particular, even when used in a referential way as in the case of the rods. Thus a student's ability to think in general terms based on selected instances may interact with the student's knowledge of the features of the notation system that can be used to express generality.

But there is a third conceptual issue when we are dealing with *mathematical* generality, because the idea of systematic variation over the containing class is involved. Freudenthal (1983) describes two different forms of numerical variation expressible algebraically. One, which he terms "polyvalent name," simply involves membership in a class, as when a letter is used to describe a general rule that applies to all particular instances of values of the letter. For example, when one writes commutativity of multiplication, $a * b = b * a$, the letters a and b are polyvalent names. The other, which he terms "variable object," involves systematic variation, with almost a kinetic aspect, as when one writes $\text{Lim}_{x \to 0} \sin(x)/x = 1$. Here x is a variable object. (See Kaput, 1979, for a discussion of the underlying motion metaphor.)

Variable objects, because they seem to supply some of the variation themselves, seem easier to understand than polyvalent names, which also came later in the history of mathematics. In the case of sum of integers, we are near the boundary of these two forms of variation, for on one hand the n in the formula $n(n + 1)/2$ originates as a variable object, a shorthand for the growing numbers at the end of the table, or as the length of the longest rod in a staircase. On the other hand, once it appears in the formula, it acts like a polyvalent name that holds for all particular instances. Thus there seems to be a difference in status according to whether one is writing the statement initially or interpreting it as an existing written notation. It seems to be *written* as a description of a particular case and then *interpreted* as a general rule.

I have examined the generality issue in some detail because it arises so often in mathematics learning and usage, and because it has aspects that can be considered in representational, read-write, terms. Here, for example, the formula in A can be regarded as representing the results of incrementing actions on both B and C. But neither the actions on B and C nor the form of their results embody the generality embodied by the algebraic notation system A, so there is an inherent mismatch between A and the other two systems. This mismatch is, I believe, quite a common source of student representational difficulty when students attempt to express generality based on the particular.

2.5 Sum of Integers and Iconicity

Often, after completing the development using the rods described above, I revisit the problem in the language that the young Gauss is assumed to have used to sum the integers from 1 to 100, say. In particular, he is assumed to have regrouped the integers, adding 1 to 100, 2 to 99, 3 to 98, and so on. Each such sum is 101, so the only question left is how many such sums are there as we work from the "outside to the middle." To help with this I ask what is the last 101 sum possible in the middle? To help further, I re-array the integers in two rows, but using a smaller set, namely the integers from 1 to 10, as in Figure 7.

1	2	3	4	5
10	9	8	7	6
11	11	11	11	11

Figure 7.

Here, quite explicitly, there are five sums of 11 each, from which students are quickly able to extend to the 100 case, to say that there will be 50 sums of 101 each, so that the final sum will be 5050. Thus we reproduced in particular terms, but not general terms, what Gauss did. There was also the matter of what to do if the integer is odd.

Here I offer the suggestion that we mimic what we did with the rods, but with numbers, namely, find double the sum and then take half the result. This leads to the array in Figure 8, exploiting a convention for expressing generality in mixed numerical and algebraic notation.

1	2	3	...	n-2	n-1	n
n	n-1	n-2	...	3	2	1
n+1	n+1	n+1	...	n+1	n+1	n+1

Figure 8.

Thus the general sum we are looking for is half of this, etc. The reader will note the strong parallel between the above sum and the array of rods pictured in Figure 6. I have manipulated the combined notation system for numerals and algebra into a form D that is, in a sense, an "iconic" version of the rods B – provided the reader (or student) interprets the alpha-numeric characters quantitatively. So now we have four systems, A (algebra), B (rods), C (traditional table of numerals), and D (specially organized hybrid numeric/algebraic sum), as in Figure 9.

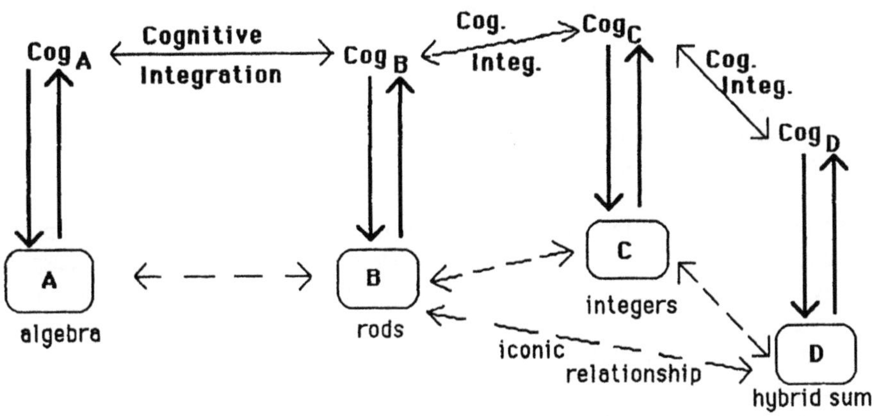

Figure 9.

An underlying question of this paper concerns the "allocation of responsibility" for the structure of reasoning processes between notation features (as described by a third-party) and the cognitive states of the person interacting with those notations. As I have attempted to make clear, there is no notational feature without an interpretive act by an interpreter, and such acts require prior knowledge on the part of the interpreter. This is what I have meant by a feature being "available" to support or constrain reasoning processes. A corollary of this statement is that iconicity is a property of a pair of notations that may or may not be available to a particular interpretant depending on the perceptual/cognitive availability of the operant features in the two notations.

I illustrate with the case at hand. The hybrid sum D and the rods B share particular features that are usually perceptually/cognitively available to the students. These features include the horizontal nature of the arrays, the ascending-descending paired horizontal sequences of items, and the vertically oriented "sums." But a more critical matter is the key role of these specific features in the reasoning processes leading to the formula. They each help identify the double sum as a repeated addition of a fixed number, and they make explicit the number of terms in this sum so that it is amenable to description as a simple product. A third important factor at work here is the borrowing (or superposition) of the notational convention from algebra A used to denote generality in the new iconic context – both use of ellipses and the literal n. In effect, we see a kind of notational opportunism at work here, where we bring together in common harness features useful to accomplishing the task from each of three different notation systems, A, B, and D. For a different perspective on the complex issue of iconicity, see Goodman (1976).

2.6 Reflections on the Sum-of-Integers Example

Dual Reciprocity: Between Cognition
And Notations, And Between Notations

We have seen several instances of different notations constraining and supporting cognitive processes, and doing so in different ways. This whole process is dually reciprocal – reciprocal in both the horizontal and the vertical dimensions of the generic layout in Figure 4. I will trace this reciprocity, drawing on the concrete example as reflected in Figure 9.

Read-Write Reciprocity for a Single Notation System: As indicated earlier, a physical notation, as with the rods B, both evokes cognitions (perhaps affect also) and serves as the object of an active interpretive process. These two processes, one cognitively (but not perceptually) passive and the other cognitively active and deliberate, constitute the building and then elaboration of Cog_B in what in itself is likely to be a cyclical process alternating between these models of construction (Grossberg, 1980). This comprises the upward pointing arrow from B to Cog_B. On the other hand, once some version of a Cog_B is constructed, students can act intelligently on the notations (rods), adjoining them, rearranging them, counting them, etc. These actions originate as mental operations at the cognitive level (producing cognitive state changes) and are projected downward in that form of "writing" which the system accepts. Such projections alter the physical state of B and provide means for comparisons (1) between states and (2) between produced and expected states as alterations make the "upward move" to the cognitive level and lead to additional processing. Such processing is followed by more projections, actions and reflective comparisons in a pattern of reciprocal influence between material notations and cognitive structures. Note that this overly brief summary does not pretend to address the huge amount of variety in such activities as notations and tasks vary and as cognitive states vary.

Reciprocity Between Notations: This reciprocity has two aspects: (1) the coordination of different cognitive structures that are brought to bear in the given situation, (2) relations between available features of different notations.

For example, in the task of summing integers, the cognitive coordination involves Cog*B* associated with the rods, probably based in sensori-motor actions and structures, and Cog*C* associated with the quantitative structure of integers brought to the students' attention by the teacher. The coordination of these two systems of cognition is one (the upper) portion of the (horizontal) dimension of the reciprocity. Similar coordination of cognitive structures must take place associated with the other pairs, e.g., between C and A, and then between D and A, with the likelihood that the structures involved are modified at each stage as a result of the coordinating acts. That is, they are not assumed as fixed cognitive structures.

The second aspect of notational reciprocity concerns experienced matches and mismatches between features of material notations. Of course, such feature-based analysis applies only to features that are perceptually and cognitively "available" in the sense defined earlier. In terms of Figure 4 one may choose a notation B in a familiar system such as natural language or a physical material that has associated with it a rich cognitive structure Cog_B that can serve as the base for constructing a Cog_A. In this way one constructs a representation of B within the notation system A. In summary, the strategy might be described as an "up-over-down" route in the diagram, beginning with B and ending at A, which is a representation of B (which in turn was an up-over-down representation of C, a table of numerals). This was my strategy in using the rods above. It was a pedagogical choice, and subsequent actions based on that choice had B representing C rather than the other way around – again, there is no assumption of "natural" or pre-existing representation here.

Figure 9 exhibits our deliberate use of (available) feature matching between double staircases and formal sums of integers, with selected features of the algebraic system superimposed on the numerals to help yield a general formula rather than a particular numerical sum. In this array of several interlinked notations, the matter of which may be used to represent which, i.e., which is used as referent and which is used as representing symbol, is a matter of how they are being viewed and used at a particular instant – a matter of one's intentions rather than any pre-existing characteristics of the notations. The rods might be used to represent the hybrid sum at the outset, as a means of interpreting it. But later, the rods come to be thought of as represented by the sum, which is a much more general mathematical statement. In fact, that particular hybrid sum comes to subsume the algebraic representation treated as separate previously – it "swallows" A. In this sense it is a very potent notation system, capable of being used to represent several other notations. Indeed, our shared notation systems vary greatly in their respective abilities to serve our representational purposes, a variation that is not yet well understood.

2.7 Notations as Cognitive Constraints and Supports

The reciprocities outlined above reveal complex interactions among notations and the cognitive structures associated with their use. I have not yet addressed in general terms how or why particular features of those notation systems actually function in reasoning processes, either in reasoning *within* a system or in translation *across* systems when they are being used in a referential way. Dealing systematically with these questions is crucial to building effective learning environments that capitalize on concrete embodiments or that build fluent use of multiple, linked notations. In work in progress I examine these questions in the context of students learning the conceptual field of multiplicative structures by using a variety of concrete notations that help support different reasoning processes.

One major pedagogical value of such concrete and (relatively) explicit systems lies in the richness and robustness of the cognitive structures that students typically have available to work with, the extent to which these are shared among students, and most importantly, *the specific kinds of cognitive processes that they evoke, support and constrain.* Notational systems vary greatly in this property. In some of our work in the Multiplicative Structures Project at the Educational Technology Center we addressed directly the practical question of how to capitalize on this kind of concretely based proficiency – how to extend reasoning with particular concrete support to reasoning with the more flexible and powerful notation systems of mathematics. We have used the strategy of deliberately linking notation systems in computer learning environments and a variety of techniques to extend reasoning processes from concrete to more abstract systems in what I call a "concrete-to-abstract software ramp" (Kaput & Pattison-Gordon, 1987; Kaput, et al., 1988).

These practical questions drive further research, however, research on how and why the particular systems operate cognitively as they do. As we have seen, some of the factors can be quite subtle, involving:

* the nature of the actions that are supported,
* the explicitness of the representation of relationships among states of the notational system (in terms of cognitively relevant physical cues),
* how the results of actions (states) are accumulated, recorded and displayed, and
* the fit between the salient available features of the system and the mathematical structures that one hopes are being cognitively constructed.

Further, some notations are used mainly for display purposes (e.g., a table of data or a coordinate graph) whereas others are designed primarily to be acted upon (algebraic expressions, arithmetic expressions). Notations also vary greatly in their "procedurality," which relates to their ability to support "runnable mental models." Two notations A and B that are being treated as referentially related, say A being used to represent B, can also vary greatly in the kinds of support and constraint that they provide. For example, one might use the conceptions based on B to guide the actions on A. I have referred to such as "semantically guided" actions as opposed to "syntactically guided" ones, where a connection between A and B is temporarily ignored, and actions performed on A are based strictly on one's knowledge of the syntactical rules that define appropriate actions on A. This distinction parallels Bruner's distinction between transparent and opaque uses of symbols (Bruner, 1973). For further discussions of these issues, see (Kaput, 1989).

2.8 What Have We Not Done?

Despite this chapter's length, I have not dealt with the affective aspects of symbol use in mathematics (Cobb, P., Yackel, E., & T. Wood, 1989; Kaput, 1989), nor have I dealt with the fundamentally social nature of symbol use (see Richards, this volume). Further, my descriptions in the sums-of-integers example have been rather course grained and idealized. Observations of individuals show much more richness in the reciprocities than the bare interactions that I have depicted. Also, I have treated notation systems as if they were all static, but some of the pedagogically more interesting situations in the future are likely to involve creation and interaction with dynamic systems. Finally, I have tended to treat notations as culturally received instruments being "given" to students rather than as individual creations developed on an ad hoc basis and refined by further use, which is closer to my preferred pedagogy. Clearly, work towards understanding the constructive role of notations has only just begun.

REFERENCES

Bruner, J. (1973) *Beyond The Information Given*, J. Anglin (ed.) NY: Norton & Co.

Cajori, F. (1929a) *A History of Mathematical Notations, Vol. 1: Notations In Elementary Mathematics*, LaSalle, IL: The Open Court Publishing Co.

— (1929b) *A History of Mathematical Notations, Vol. 2: Notations Mainly In Higher Mathematics*, La Salle, IL: The Open Court Publishing Co.

Clement, J. (1982) "Algebra Word Problem Solutions: Thought Processes Underlying A Common Misconception." *Journal For Research in Mathematics Education. 13*, 16–30.

Cobb, P,. E. Yackel, and T. Wood (1989) "Young Children's Emotional Acts While Doing Mathematical Problem Solving." In D. McLeod & V. Adams (eds.) *Affect and Mathematical Problem Solving: A New Perspective.* NY: Springer-Verlag.

Fischbein, E., D. Tirosh, and P. Hess (1979) "The Intuition Of Infinity." *Educational Studies In Mathematics*, 10, 3–10.

Freudenthal, H. (1983) *Didactical Phenomenology Of Mathematical Structures*, Boston: D. Reidel.

Goodman, N. (1976) *Languages Of Art.* (Revised ed.) Amherst, MA: University of Massachusetts Press.

— (1984) *Of Mind And Other Matters.* Cambridge, MA: Harvard University Press.

Greeno, J. (1988) "The Situated Activities Of Learning And Knowing Mathematics." In M. Behr, C. Lacampagne, M.M. Wheeler (eds.) *Proceedings of the 10th Annual Meeting of the PME-NA.* DeKalb, IL.

Grossberg, S. (1980) "How Does A Brain Build A Cognitive Code?" *Psychological Review*,

87. 1–51.

Hofstadter, D.R. *Gödel, Escher, Bach: An Eternal Golden Braid*, NY: Basic Books, 1980.

Hooper, C. (1986) "Architectural Design: An Analogy." In D. Norman and S. Draper (eds.) *User Centered System Design*. Hillsdale, NJ: Erlbaum.

Kaput, J. (1979) "Mathematics and Learning: The Roots Of Epistemological Status." In J. Clement and J. Lochhead (eds.) *Cognitive Process Instruction*, Philadelphia: The Franklin Institute Press.

— (1985) "Representation and Problem Solving: Methodological Issues Related To Modeling." In E. Silver (ed.) *Teaching and Learning Mathematical Problem Solving: Multiple Research Perspective*, Hillsdale, NJ: Erlbaum.

— (1987a) "Representation and Mathematics." In C. Janvier (ed.) *Problems of Representation In Mathematics Learning and Problem Solving*. Hillsdale, NJ: Erlbaum.

— (1987b) "Toward A Theory Of Symbol Use In Mathematics." In C. Janvier (ed.) *Problems of Representation in Mathematics Learning and Problem Solving*. Hillsdale, NJ: Erlbaum.

— (1989) "Linking Representations in the Symbol System of Algebra." In C. Kieran & S. Wagner (eds.) *A Research Agenda For The Teaching And Learning Of Algebra*. Reston, VA: National Council of Teachers of Mathematics; and Hillsdale, NJ: Erlbaum.

Kaput, J. and L. Pattison-Gordon (1987) "A Concrete-To Abstract Software Ramp: Environments For Learning Multiplication, Division, and Intensive Quantity." Technical Report 87–8, Educational Technology Center, Harvard Graduate School of Education, Cambridge, MA 02138.

Kaput, J., M.M. Katz, C. Luke and L. Pattison-Gordon (1988) "Concrete Representations For Ratio Reasoning." In M. Behr, C. Lacampagne, M.M. Wheeler (eds.) *Proceedings Of The 10th Annual Meeting of the PME-NA*. DeKalb, IL.

Morrow, D.G. (1986) "Grammatical Morphemes And Conceptual Structure In Discourse Processing." *Cognitive Science*, 10 (4) 423–55.

Peirce, C.S. (1933) *Collected Papers*. Vols 1 and 2. Cambridge, MA: Harvard University Press.

Popper, K. (1972) *Objective Knowledge*. Oxford: Oxford University Press.

Steffe, L.P., P. Cobb and E. von Glasersfeld (1988). *Construction Of Arithmetical Meanings And Strategies*. NY: Springer-Verlag.

Toulmin, S. (1972) *Human Understanding*, Vol. 1, Princeton: NJ: Princeton University Press.

— (1982) *The Return To Cosmology*. Berkeley, CA: University of California Press.

Vergnaud, G. (1987) "Conclusion." In C. Janvier (ed.) *Problems of Representation in Mathematics Learning and Problem Solving*, Hillsdale, NJ: Erlbaum.

von Glasersfeld, E. (1987) "Preliminaries to any Theory of Representation". In C. Janvier (eds.) *Problems Of Representation In Mathematics Learning And Problem Solving*, Hillsdale, NJ: Erlbaum.

Werner, H. & B. Kaplan (1962) *Symbol Formation*. NY: Wiley.

Wheeler, M.M. & W.G. Martin. "Explicit Knowledge of Infinity." In M. Behr, C. Lacampagne, M.M. Wheeler (eds.) *Proceedings Of The 10th Annual Meeting Of The PME-NA*. DeKalb, IL.

MAKING MATH MEAN

JACK LOCHHEAD

University of Massachusetts, Amherst, MA

In this chapter, I will attempt to describe some of the implications of constructivism that have proved useful in the establishment and maintenance of a moderately large mathematics course. That is, I will not be considering some theoretical ideal, but rather describing what has proved possible under relatively normal conditions. The course in question has over a dozen sections, as many instructors, and 30–40 students per class. The course is designed to help students remedy deficiencies in their previous mathematics preparation and therefore attracts a very diverse clientele.

Constructivism has implications both for *what* we teach and for *how* we teach it. First, constructivism is a statement about the nature of knowledge and its functional value to us. Mathematical knowledge viewed from this perspective is not the same as when viewed from other perspectives. Second, constructivism implies a mechanism for how we acquire knowledge and, hence, how it is possible to teach. To date constructivist thinking has been more effective in describing what sorts of teaching will not work than in specifying what will. But it does at least define a direction for future exploration.

Finally, constructivism implies new relationships between the teacher, the learner, and the content being studied. These new relationships often do not fit the expectations of the students, the teachers, and the teachers' supervisors. Even without using an extreme fully constructivist approach teachers often provoke reactions such as the following:

> If teachers would stick more to the facts and do less theorizing, one could get more out of their classes .... A certain amount of theory is good, but it should not be dominant .... The facts are what's *there*. And I think that would be ... the main thing. (Perry, 1970 p. 67).

In the mathematics classroom the conflict can be even more severe. As Lakatos (1976) and Klein (1980) suggest, mathematics is the last refuge of those who believe in certain knowledge, and to challenge the absolutes of mathematics is to shake the foundations of all truth. No self respecting student can let this go by without at least a whimper.

1.1 What to Teach

Knowledge is used to organize past experience and to predict future ex-

E. von Glasersfeld (ed.), Radical Constructivism in Mathematics Education, 75–87.

periences. There is no unique or even preferred way to do this. Another person's knowledge can be thought of as equivalent to our own only insofar as it seems to induce them to react to experiences in ways we would predict. Thus the teaching of mathematical content needs to be understood in terms of a set of mathematical experiences which the teacher attempts to arrange for the student. When the teacher believes that s/he and the student have negotiated a consensual domain (Maturana, 1978) within the realm of these experiences, then the material can be said to have been taught. The process is every bit as slippery as the description is convoluted. In short, mathematics is taught by having students do mathematics. When students react appropriately to a set of mathematical situations then we can assume they have learned a mathematics for those situations. We cannot automatically assume that their mathematics will lead them to react to new situations in the same way we do. Therefore it is essential that we are careful to pick a set of mathematical experiences that includes a basis for all the mathematics we want to teach. Since most mathematics is taught for the purpose of enabling students to handle certain specific applications it is important to use those applications in teaching, and to consider very carefully the range of applications that will be explored. None of this is revolutionary although it can be highly controversial. Rather, there is a somewhat subtle shift in emphasis giving greater importance to examples and especially to the thought that goes into their selection.

Constructivism implies a rather convoluted approach to the teaching of mathematical content. It is much more straightforward on the teaching of mathematical thinking. Here the emphasis is on teaching students to be more effective constructors of mathematical concepts. Three aspects are worth special mention.

1.2 Relating Mathematical Knowledge to Other Knowledge

At the level of pure mathematics, as experienced by pure mathematicians, there is no need for reference to the nonmathematical world. But the experience of these individuals is based on a complex mathematical machinery which they have constructed over a period of years. The mathematical novice must construct his/her knowledge using conceptual structures and experiences which have already been constructed from their own, largely nonmathematical, experience. Students need to learn to evaluate their mathematical constructions in relation to their other knowledge. Since normal school mathematics often has the effect of discouraging such comparisons, it can be quite difficult to convince university students of the relevance of such activity.

It is also difficult to pick examples that really fit the students' experience. For example, one may try to evaluate knowledge of the weighted mean by comparison to the typical American college and high school grading system. A simple question would be to calculate the grade point average of 6 credits of A in French with 3 credits of F in German. Unfortunately research has shown

(Hardiman, 1983) that although most American students are subjected to this sort of grading system from at least the age of 13, few college students actually understand it. In one study a student who had proposed the unweighted mean, $A(4.0) + F(0.0)/2 = 2.0$, was asked if it was fair to count the 3 credit course as much as the 6 credits. He replied "it is not fair but that's how they do it."

1.3 Checking for "Self-Consistency"

A more traditional method of evaluating mathematical constructions is checking for self-consistency. However, consistency really is only in the eye of the beholder, Lochhead (1988). A striking example comes from the extension of the concept of exponent beyond whole numbers. In deciding what meaning to associate with a negative exponent one can investigate what interpretation would be consistent with previous conventions, and in particular the observation that

$$x^a/x^b = x^{a-b}$$

Here self-consistency implies that a negative exponent should be interpreted as the multiplicative inverse, $x^{-a} = 1/x^a$. But this interpretation comes from a selective notion of consistency because it ignores the original definition in which the superscript was the number of times the base number was to be multiplied by itself, and there is no way to multiply a negative number of times. It is even worse with fractional exponents, as in $x^{1/2}$. Students therefore, often find such definitions by consistency to be very suspect. To overcome such concerns it is desirable to expose students to quite a few examples of this kind of reasoning and to point out the inconsistency in the consistencies sought.

1.4 Think Before You Fact

Perhaps the most poisonous piece of realist dogma corrupting current instruction is the notion that students need to be taught some set of basic facts before they can be asked to think. This misconception springs from a failure to understand the dynamic nature of facts. Neither facts nor the process of learning them are static. A fact is the output of a thinking process and the fact production system is itself constructed via thinking. Thinking must come first. When a set of "facts" are memorized prior to serious work, that merely means that little thought has gone into their construction and that they are therefore ill-defined and poorly constructed. Knowledge composed of such facts tends to be fragile, disorganized and difficult to apply. While it is certainly possible to begin learning by acquiring large numbers of fragile facts which are later laboriously reshaped into coherent knowledge structures, it is not obvious that this is the most efficient way to proceed.

The reason that the dogma of putting facts before thinking survived as long as it has, probably stems from the difficulty teachers have in conceiving thinking activities that are appropriate to the learner's situation. To better understand that difficulty it is useful to consider the child's construction of number as described by Steffe, von Glasersfeld, Richards and Cobb (1983). Here it is shown that children first learn to recite the number words in sequence prior to understanding counting or number. It is tempting to assume that before mastery of the number word sequence there is little opportunity for numerical thought. That is not the case. Tasks such as distinguishing number patterns or matching up items as in setting a dinner table can be quite thought provoking. Instruction that emphasizes such activities can be remarkably effective, e.g. see Paul Cobb's chapter in this book. Thus even the most mundane knowledge should be seen not as a set of facts to be learned, but as a set of situations to explore.

1.5 How We Teach

The most important characteristics of constructivist teaching are skepticism and curiosity. One should always remain skeptical about the effects of one's own efforts as a teacher and curious about the efforts of one's students. In general the teacher should spend more time listening to students than the students spend listening to the teacher. But before a teacher can be convinced of the need to reverse the usual ratio of listening times s/he must recognize just how ineffective certain traditional forms of telling tend to be.

Lectures can be a great deal of fun, especially for the lecturer; they can be inspiring and occasionally thought provoking, but they are rarely effective for producing polished knowledge. One way to convince oneself of the limits of lectures is to try the following type of experiment. Deliver a lecture on some non trivial but relatively simple concept. Leave enough time at the end to ask students to describe the concept in their own words; do this by asking a question which makes sure students will not respond by simply repeating your own words.

Let me share with you two of my own attempts. The first comes from an introductory physics class and was given about midway through the semester. The following are students' answers to a quiz question:

The question was asked directly after a lecture on force in which the "force causes velocity" misconception was addressed directly! Students had been explicitly told that the correct equation was Force = mass times the *change* in velocity per unit time ($F = m\,dv/dt$), and that force is *not* proportional to velocity (i.e., $F = kv$, is false). These equations were explained with examples and informal verbal descriptions as well as the standard formal explanations.

The question I asked was: "What is the cause-and-effect relationship between force and velocity?"

I got the following student answers:

"As one increases, the other increases, and as one decreases, the other decreases."
"A change in force causes a change in velocity."
"Force causes velocity to occur. So if force increases, so does velocity, be it negative or positive."
"The stronger the force on an object, usually the velocity is greater".
"When the force increases, velocity increases, and when force decreases, velocity decreases. In other words, force is the cause and velocity the effect."
"If the force is constant, the velocity will remain constant. If the force is increasing, velocity will increase. If the force is decreasing, the velocity will decrease."
"Force is the cause which gives the object its velocity."
"Force causes velocity, but when force stops, the effect may go two different ways: it may either speed up or slow down."

To the nonphysicist it may not be apparent that these answers miss the mark. In fact, making a reasonable guess as to what each answer meant to its author is far from trivial. This is an illustration of why it is important for the constructivist teacher to know her/his subject well, even though he/she may not choose to expound upon it.

Another example comes from a course in calculus.

I asked the following question halfway through a semester:

"During the past 6 months, the rate of inflation has dropped from 18% to 10%. Explain what this change means in terms of the value of the dollar."

Of the 26 responses, 16 stated that the value of the dollar would increase. The ten that were not clearly wrong are given below.

* * * * *

"In terms of the dollar if inflation has dropped the value of the dollar had increased or at least its decrease in value has slowed down."

* * * * *

"The value of the dollar is decreasing with inflation, but if the rate of inflation drops, then the change in the value of the dollar is increasing."

* * * * *

"Dollar increasing 18%
10% increase Decrease of 8%
* The value decreases by 8%."

* * * * *

"When the rate of inflation is large, the value of the dollar goes down because what you used to be able to buy with a dollar you can't anymore because they cost more. If the rate drops, then you should be able to buy more with your dollar after. Everything will cost more, but it will just take a longer time to get to the price, it would have with a higher rate of inflation."

* * * * *

"t = 6 months
value has gone from 18% to 10%, .18 on a dollar to .10 on a dollar. di/dx = dropped 8%.
Dollar will be worth less as the months go on."

(This answer was accompanied by a large graph showing a downward curve of dollar value, with inflation on the y axis and months on the x axis.)

* * * * *

"This *change* means the *value* of the dollar is still being inflated, but not as fast - the slope of the curve measuring the rate of inflation is still positive but decreasing."

* * * * *

"The rate of inflation has dropped from 18% to 10%, a decrease in 8%. All this means is that the *rate* of inflation has slowed down, not that there is no inflation. The value of the dollar will not really change."

* * * * *

"6 months
rate of inflation – 18% – 10%
Value of a dollar = ?"

The change of the inflation rate from 18% to 10% means that the value of the dollar is not going to decrease so rapidly. This means that the increase of the inflation rate has slowed down.

This can be seen by plotting the two percent's as slopes of .18 and .10. The slope of .10 does not increase as rapidly, rate of inflation.

The 2nd derivative would tell how the slope (1st derivative) is changing – change in the value of the dollar."

* * * * *

"t = time
v = value of $ I = inflation

Change in value = dv/dh rate of inflation = dI/dH

This change means that as the rate of inflation drops, the value of the dollar is dropping, as time increases."

(This answer also has a graph plotting time along the x axis and value of the dollar on the y axis.)

* * * * *

Once again it is apparent that questions of this type produce answers which can only be accurately interpreted by a teacher with a solid understanding of his/her subject. Of course struggling with such answers is a very good way to sharpen one's own understanding and is invigorating as well as difficult.

1.6 Tutoring

Tutoring is about as effective as lecture unless a great deal of effort is made by the teacher to find out what the student is really thinking and the teacher is also able to formulate situations that will induce the student to question inadequate concepts. Rosnick and Clement (1980) describe a series of tutoring studies that illustrate just how ineffective even carefully planned tutoring can be. One particularly successful student (in terms of right answers), later explained his strategy as figuring out what made sense and then doing the opposite. This case high-lights the point that both lecture and tutoring can appear to be highly effective if they are evaluated in terms of narrowly defined behavioral objectives. It is only after one has probed for a deeper level of understanding that these modes of teaching become suspect.

1.7 Telling It Like It Is

We often tend to operate under the naive belief that notation carries its own interpretation. The careful and consistent use of conventional notation is an important part of good instruction, but it is by no means an adequate substitute for dealing with the underlying mathematical concepts. It is nearly impossible for a person familiar with the conventions of mathematics to see the extent to which the system is inconsistent and illogical. Careful attention to the struggles of neophyte students quickly reveals numerous quirks. The operation of dividing 4 by 3 is written as 4/3 or $4 \div 3$ but also as 3 into 4, so much for the preservation of order. Most mathematicians would interpret $f(x + 1)$ as meaning some function of $x + 1$ but would recognize $c(x + 1)$ as c times $x + 1$. One of the most dramatic demonstrations of the inability of notation to convey its own meaning was a study by Rosnick (1981) in which he asked the following question:

At this university, there are six times as many students as professors. This fact is represented by the equation $S = 6P$.

A) In this equation, what does the letter P stand for?
- i) Professors
- ii) Professor
- iii) Number of Professors
- iv) None of the above
- v) More than one of the above (if so, indicate which ones)
- vi) Don't know

B) What does the letter S stand for?
- i) Professors
- ii) Student
- iii) Students
- iv) Number of students
- v) None of the above
- vi) More than one of the above (if so, indicate which ones)
- vii) Don't know

Over 20% of calculus graduates chose S stands for professor, and this was not a random guess since every one of these students picked "none of the above" in answer to part A.

In spite of all the evidence against it, mathematics teachers keep searching for the self interpretive notation. Many believe that the confusion in the above problem would vanish if Ns and Np were used instead of the variables s and p (e.g., Fischer, 1988). The letter N is believed to convey the idea that variables are numbers, but there seems to be no need to tell students how or why it is being used.

The above examples of the failures of lecturing, tutoring and careful notation were given not to belittle these important elements of an effective instructional system but rather to place them in fairer context with several other more intuitively suspect components. These other components involve situations in which students are in some degree left to their own devices and in which teachers are unable to keep track of all of the ideas which the students are constructing. (*This of course is always the case* but in conventional instruction it is easier for the teacher to ignore the students' ideas). To remain sane in these situations, teachers need to cultivate the attitude of trusting the students' own efforts to understand the material, not in the sense of believing that the students' creations will be reflections of the "accepted truth" but in the sense that they are all there is to work with. The more carefully the students think about their ideas the better those ideas will become; it is of little direct help if the teacher thinks carefully about them. (Of course it is enormously important for the teacher to *have* thought carefully about the topic in question, otherwise it will be impossible to make much sense out of the students' creations.)

One powerfully constructivist mode of instruction is student discussion, often in the form of some type of cooperative problem solving, with or without the supervision of an instructor. The primary goal of such discussions is to help students develop skills of constructing, evaluating and modifying concepts in the domain of interest (in our case mathematics). The teacher's role therefore is to work to improve the quality of the discussions rather than to focus from the beginning on the "correct" mathematical answer. I recently worked with a student who was confronted with the problem of finding a number which, when multiplied by 6, would yield 3. He began by asking himself whether that number would be greater or less than 10. On discovering that 6 times 10 yielded 60 he decided to try numbers less than 10 and gradually worked his way down to 6 times 1. Since this was still too big he was puzzled as to what to try next. Telling him that the answer was one half provided no transfer to finding a number which when multiplied by 9 would yield 3. While I was concerned about the student's lack of number sense, I was more impressed by his analytical reasoning. A short time later I found this same student playing a significant role in student discussion of a very difficult word problem involving fractions. By the end of the semester he was one of the best students in his class. The point is that the facts will straighten themselves out once the tools for analyzing them are developed and refined.

We have found that our best instructors spend a great deal of time listening to student discussions (which are the major component of our classes) but very little time actively participating; but when they do contribute it is in such a way as to leave the students in control and to insure that all student remain active participants. To restrain one's natural enthusiasm to such a limited (but highly effective)role, it is necessary to have developed a great deal of humility concerning the probable impact of one's words of wisdom. It is also critical to appreciate the value in having a well developed sense of how the students conceive the mathematics.

One method for obtaining insight into the student's mathematical thinking, while at the same time giving students the opportunity to reflect on their own ideas, is to assign thought process protocols. In such an assignment, students are to write down everything they are thinking as they work on a problem. This is far from easy to do. An example is given below.

PROBLEM STATEMENT:

Flask Problem

A flask containing 650 doses of medicine had 5 ml. of the medicine removed for experimentation. Because of an emergency, another clinic was sent one-fifth of the remainder. Afterward, only 512 doses remained. Calculate the dosage in milliliters (ml.). Hint: Be careful to recognize that there's a difference between number of doses, and number of milliliters per dose.

PROBLEM SOLUTION:

I've read the problem a couple of times, and I'm not sure where to start. "A flask containing 650 doses of medicine ..." – why don't I start by drawing a flask?

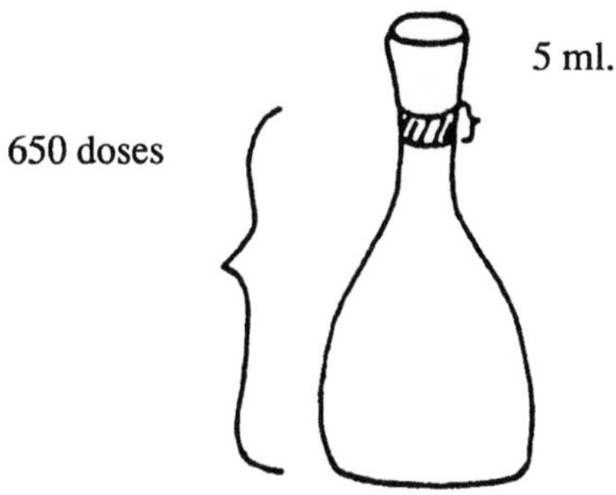

Here is a full flask and it contains 650 doses.

Next, 5 ml. are removed. I don't know how much that is in proportion to the 650 doses, so I will just shade in a little bit at the top which means it has been taken out.

Now, another clinic was sent one-fifth of the remainder. That means that I will have to divide what is left in the flask into 5 equal parts. But, how do I do that? The flask is a funny shape, and if I draw lines to separate it into 5 parts, I won't be sure that they will be 5 equal parts. Is there another way I can draw the picture? Does the flask have to look like the kind in the chemistry lab, or can I just use a box to represent 650 doses? I don't see why not. After all, 650 doses is 650 doses – right? I think so.

O.K. I'm starting over again. I'm going to draw a box which represents 650 doses of medicine. And, I'm going to take 5 ml. out of one corner.

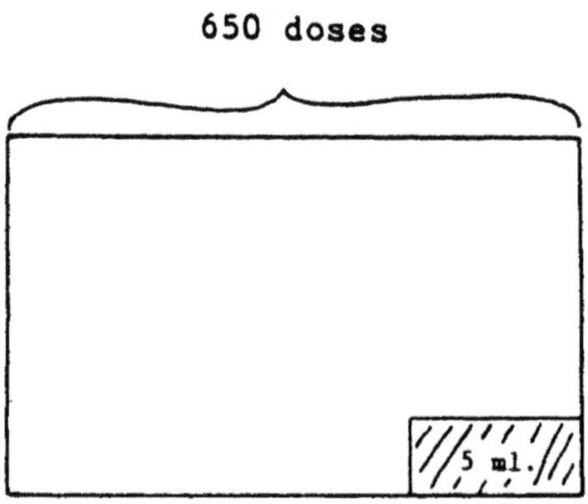

I shaded it to show that it was taken out.

Oh, no. I have the same problem as before. I can't separate the remainder (after I take the 5 ml. out) into 5 equal parts. I have to do it another way. I will take the 5 ml. out of one end.

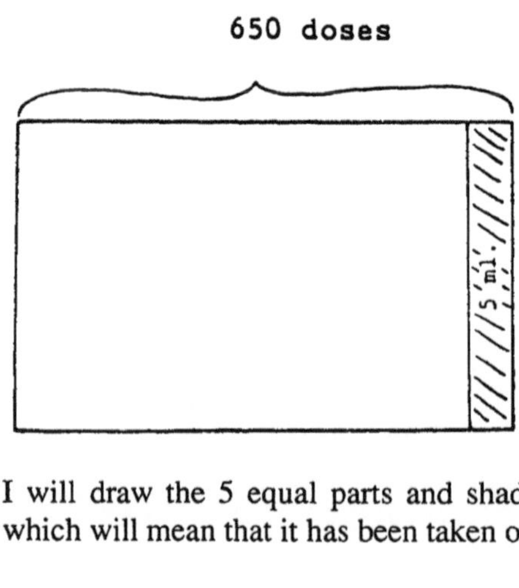

Good. Now I am left with something I can section into 5 equal parts.

I will draw the 5 equal parts and shade in the portion sent to another clinic which will mean that it has been taken out.

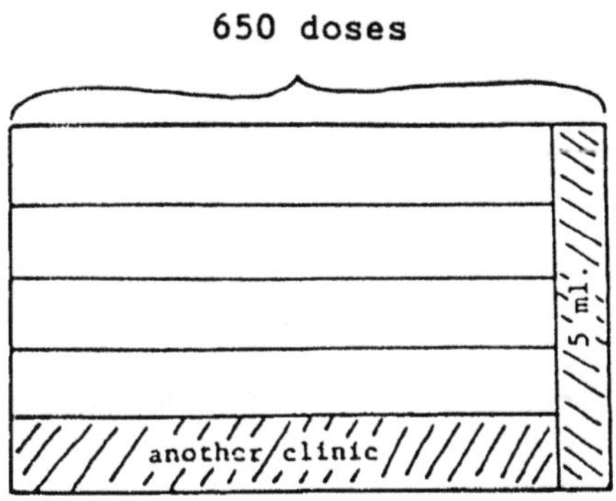

Now, the problem says that only 512 doses remained. That means that there are 4 equal parts remaining, which total 512 doses. Ah ha. Then, there are 512 - 4 = 128 doses in each part. That must also mean that the other clinic was sent 128 doses since their portion is the same size as each of the other equal parts. I will write this information in my drawing.

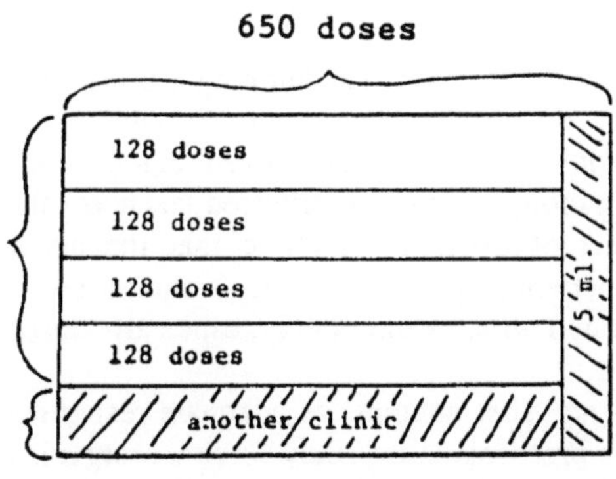

Let me review what the drawing tells me. There are 650 doses total. 512 doses are still in the flask and 128 doses were sent to another clinic. Then 512 + 128 = 640 doses accounted for. What about the 5 ml.?

There are 650 doses in all. 640 of them are accounted for. Then, 10 remain. Those 10 doses must be 5 ml.

If there are 5 ml. in 10 doses, how many ml. in one dose? This reminds me of one of those "miles per gallon" problems. Only now it is "milliliters per dose." With "miles per gallon", I always put the miles on the top and the gallons on the bottom like this:

$$\frac{\text{miles}}{\text{gallons}}$$

I will set up the "milliliters per dose" the same way:

$$\frac{\text{milliliters}}{\text{dose}} = \frac{5}{10}$$

If that is reduced, it comes to 1/2. That means there is 1 milliliter for every 2 doses. So, if there is 1 milliliter in every 2 doses, there must be 1/2 milliliter for every one dose.

My answer for this problem is that there is 1/2 ml. of medicine for every dose of medicine in the flask.

Note: This solution demonstrates *one* way to solve this problem. There are many other ways to also arrive at the correct answer. The specific operations used in the solution may not apply directly to other problems in this section. This problem solution is intended to be a model thought process protocol. It is *not* intended to be memorized for future quizzes, tests, etc. Each problem is unique and must be approached as such.

It is important for protocols to include the informal thinking we do when confronting a real problem and often do not do when racing through a well automated exercise. An algebra problem that can be solved algebraically usually involves very little interesting thought. Thus experienced teachers often make poor role models for this sort of assignment. (Note that the above example was written by a highly experienced protocol writer, it should not be anticipated that other undergraduates will write as thoroughly, especially during their first attempts).

It is important not to accept students' first attempts as adequate responses. The virtue in writing thought process protocols is that they force a reorganization of thinking strategies. It takes a lot of frustration and struggle to reach a stage where writing protocols becomes a useful tool. Yet for those students who have the patience to master them, protocols can be exam savers. When faced with a problem that seems impossible, writing about the confusion often creates a sudden insight as to what to do. This may be for no deeper reason than that writing allows one to continue to act on the problem without becoming panicked.

1.8 Summary

Constructivist teaching involves giving up the notion that you can do for students what, in practice, they must do for themselves. It demands that you trust your students' minds as much as your own and that you have faith in people's ability to learn. It requires humility concerning one's own ability to explain or expound. But the essence of constructivist teaching is in the realization that you will never know what is going on (in the minds of your students) yet it is fun (and rather useful) to try to find out. An occasional quiz of the sort illustrated for physics and calculus, can provide teachers with a brief glimpse of whom they are teaching. A detailed probing interview offers a somewhat larger picture, but of only one student. The opportunities to explore student understanding are unlimited and unending. The investigation may be infinite, but for those involved with it, teaching need never be dull or routine, no matter how mundane the subject matter.

REFERENCES

Fisher, Kathleen M. (1988). "The Students-and-Professors Problem Revisited". *Journal for Research in Mathematics Education*, 19(3), 260–2.

Hardiman, P.T. (1983). "The Importance of Analog Knowledge in Understanding The Mean." M.S., Psychology, University of Massachusetts, Amherst.

Kline, (1980). *Mathematics: The Loss of Certainty*. New York, NY: Oxford University Press.

Lakatos, I. (1976) *Proofs and Refutations: The logic of mathematical discovery*. (Worrall, J. & E. Zahar, eds.) New York, N.Y.: Cambridge University Press.

Lochhead, J. (1988). "Some Pieces of the Puzzle". In G. Forman, and P. Pufall (eds.), *Constructivism in the Computer Age*. Washington, DC: Piaget Society Publications, 71–81.

Maturana, H. (1978). "Biology of Language: The Epistemology of Reality." In G. Miller and E. Lenneberg (eds.), *Psychology and Biology of Language and Thought: Essays in Honor of Eric Lenneberg*, pp. 27–63. New York, NY: Academic Press.

Narode, R., D. Schifter, and J. Lochhead (1985). A course guide to Math 010L. Technical Report/Curriculum, Scientific Reasoning Research Institute, University of Massachusetts, Amherst.

Perry, W. (1970). *Forms of Ethical and Intellectual Development in the College Years*. New York, NY: Holt, Rinehart & Winston, Inc.

Rosnick, P. and J. Clement, (1980, Autumn) "Learning Without Understanding: The Effect of Tutoring Strategies on Algebra Misconceptions". *Journal of Mathematical Behavior*, 3(1), 3–27.

Rosnick, P. (1981). "Some Misconceptions Concerning The Concept Of Variable". *The Mathematics Teacher*, 74(6), 418–21.

von Glasersfeld, E., J. Richards and L. Steffe, (1983). "Perspectives And Summary." In L. Steffe, E. von Glasersfeld, J. Richards and G. Cobb (eds.), *Children's Counting Types: Philosophy, Theory and Application*. New York, NY: Praeger, 1983.

TREATMENT OF REFUTATIONS: ASPECTS OF THE COMPLEXITY OF A CONSTRUCTIVIST APPROACH TO MATHEMATICS LEARNING*

NICOLAS BALACHEFF

France

INTRODUCTION

One of the basic hypotheses of research in mathematics education, one being widely considered at the present time, is the constructivist hypothesis which asserts that the subject explores actively its environment, that he or she participates actively in the creation of space, time and causality (Inhelder and Caprona, 1985, p. 8). Following this hypothesis, we can see that the student himself participates actively in the construction of his own mathematical knowledge. The starting point for the developmental process, according to this view, is the experience of a contradiction which is likely to provide a cognitive disequilibrium: It is the overcoming of such a contradiction which results in new constructions (Piaget, 1975).

I would like to emphasize here the consistency of this Piagetian developmental model with the model proposed by Lakatos to describe the growth of mathematical knowledge through the dialectic of proofs and refutations. "Informal, quasi-empirical, mathematics does not grow through a monotonous increase of the number of undubitably established theorems but through the incessant improvement of guesses by speculation and criticism, by the logic of proofs and refutations" (Lakatos 1976, p. 5). But what Lakatos' work shows is the complexity of overcoming a contradiction in mathematics due to the diversity of ways of dealing with a refutation. The historical study presented by Lakatos shows the importance of the social dimension of this dialectic. This dimension also plays an essential role in the learning process taking place in the mathematics classroom. It appears at two levels:

(i) Students have to learn mathematics as social knowledge; they are not free to choose the meanings they construct. These meanings must not only be efficient in solving problems, but they must also be coherent with those socially recognized. This condition is necessary for the future participation of students as adults in social activities.

(ii) After the first few steps, mathematics can no longer be learned by means of interactions with a physical environment, but requires the confrontation of the student's cognitive model with that of other students or of the teacher, in the context of a given mathematical activity. Especially in dealing with refutations,

E. von Glasersfeld (ed.), Radical Constructivism in Mathematics Education, 89–110.

the relevance of overcoming is what is at stake in the confrontation of two students' understandings of a problem and its mathematical content.

I have studied the complexity of students' treatment of a refutation by means of an experimental approach within the context of solving a given mathematical problem. My analysis of the data collected takes into account both the cognitive and social dimensions of the phenomena observed. Before presenting this study and its results, I would like to explain the main features of the theoretical framework I adopted.

I. THE PROBLEM OF CONTRADICTION

1.1 Conditions for the Awareness of a Contradiction

Following Grize (1983), I consider that a contradiction exists only if it has a witness. That means that a contradiction does not exist by itself, but only with reference to a cognitive system. For instance, a contradiction could be recognized by the teacher in the mathematics classroom, but ignored by students. Here is an example: In a research study about students' conceptions of the convergence of numerical sequences, Robert (1982) reports the following reasoning:

Having shown that a numerical sequence (u_n) has a limit λ such that $\lambda < 1$, some students make the following computation:

$$\lim (u_n^{n+1} + u^n_n + u_{n}2 + u_n + 1) = \lim (u_n^2 + u_n + 1) = \lambda^2 + \lambda + 1$$

They treat n in two different ways according to whether it appears as a suffix or as an exponent.

In this example the students are not aware of the contradiction which results from the two different treatments of n, whereas it is quite obvious to us.

But, on the other hand, a contradiction could exist for the learner where none exists for the teacher, for example (Balacheff, 1988):

Students of the seventh grade see a contradiction in the fact that the sum of the angles of a triangle does not depend on its size (although area or perimeter do).

In all these examples, the fact that a contradiction is elicited or not depends basically on the knowledge of the one who claims that a contradiction exists or does not.

In the teaching situation, the existence of a referent knowledge (the scientific knowledge or the knowledge to be taught) gives the right to the teacher to decide whether a fact is contradictory or not with respect to this knowledge. The problem for the teacher, then, is to make the student recognize the contradiction he or she alone sees. The didactical question thus becomes twofold: First, *what are the conditions necessary to engender, on the part of the student, awareness of a contradiction* and, second, *what are the conditions under which the student can resolve it?*

According to Piaget (1974, p. 161) the awareness of a contradiction is only possible at the level at which the subject becomes able to overcome it. I would like to emphasize that, to the extent that such an overcoming is likely to imply a real re-organization of the subject's knowledge, this condition appears too strong to us. I will not ask for such a potential overcoming of contradictions, but merely that they be seen. Actually, to become aware of a contradiction means to become able to raise the question of the choice between two propositions: an assertion and its negation. Whatever the choice, it implies that the formulation of the assertion is available and that the subject can construct and express its negation.

Thus, consciousness of a contradiction depends on two constructions, and the subject can only be aware of it when he becomes able to carry out these two constructions. But even when a contradiction has been identified, overcoming it might only be possible after a long process. For example, in the case of the sum of the angles of a triangle, overcoming the contradiction I have mentioned above, needs the construction of an invariant, and finally the conceptualization of the Euclidean Postulate and its consequences. That is to say, to construct the relation between the assertion «the sum of the angles of a triangle is 180°» and the assertion «the bigger the triangle, the larger is the sum of its angles», even if it provides access to the problem of the contradiction, is not by itself sufficient to allow it to be overcome.

On the other hand, a contradiction exists only with reference to a disappointed expectation, or with reference to a refuted conjecture. The potential existence of an assertion is not sufficient, to use a metaphor of Taine,[1] it is necessary that it comes to the forefront of the scene. Piaget himself remarks that to become aware of a contradiction is far easier when it appears between an expectation and an external event which contradicts it (*ibid.*). Here Piaget goes beyond the cognitive conditions and takes into account conditions related to the situation within which the subject acts. The existence of such an expectation means that the subject is actually committed to an assertion and is able to sidestep its action. In other words, he or she is able to consider it as a possible object of thinking, even more, an object of discourse. At that point, action is no longer just carried out. Being the product of a choice, it is considered in terms of its validity and the adequateness of its effect. That is to say, action is then related to its aim and a contradiction becomes apparent when this aim is not fulfilled. The choice which has been made and the conditions under which the action has been performed are then put in doubt.

I would like then to state the following conditions as being necessary conditions for the emergence of the awareness of a contradiction:

(i) The existence of an expectation or an anticipation

(ii) The possibility to construct and formulate an assertion related to this expectation, and the possibility to state its negation.

1.2 Treatment of Refutations

A counterexample in the mathematics classroom is often understood as a sort of catastrophe because it implies the definitive rejection of what has been refuted. From this point of view, the mathematics classroom ideology is more Manichaean than dialectic. An analysis of the mathematician's activity shows quite a different and much less radical procedure.

Taking as a basis the model proposed by Lakatos (1976), we can differentiate the implications of a counterexample depending on whether what is considered is the conjecture, its proof,[2] or the related knowledge and the rationality of the problem-solver himself.

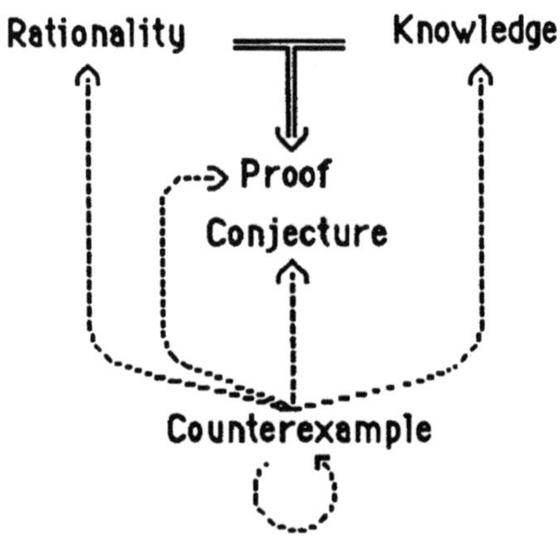

Figure 1.

The schema[3] (Figure 1), showing the conjecture and its proof as the product of both the knowledge and the rationality of a subject, summarizes the main possible consequences of a counterexample. Of course, the conjecture could be rejected, but other issues are possible. For example: To inspect carefully the proof in order to elicit a possible suspect lemma and then to incorporate it as a condition in the statement of the conjecture, or to question the rationale of the proof or the underlying knowledge; or even to dismiss the counterexample as a monster. Also, such a process is likely to provide an evolution of the knowledge related to the proof, resulting in what Lakatos called a proof-generated concept (Lakatos, 1976, pp. 88 sqq.).

The analysis Lakatos (1976) provides of the development of mathematical knowledge leaves out a question that is essential for the teacher and the educational researcher: *What determines the choice of a way to overcome the contradiction brought forth by a counterexample?* It is this question that I have investigated by means of an experimental study I will now present.

II. STUDENTS' TREATMENT OF A COUNTEREXAMPLE

2.1 *The Experimental Setting: A Social Interaction about the Number of Diagonals of a Polygon*

In order to explore the students' behavior when faced with a counterexample, I have used a situation involving social interactions which encouraged the confrontation of different viewpoints about the solution of the problem, and hence, a verbal exchange making explicit the possible refutations, how to deal with them and thus the proving process underlying the solutions proposed by the students.

Pairs of 13–14 year-old students were required to solve the following problem: *Give a way of calculating the number of diagonals of a polygon once the number of its vertices is known.* The answer to this question was to be expressed in a message addressed to and to be used by other 13–14 year-old students.

Observations, which began in 1981, were carried out mainly during the first semester of 1982. Fourteen pairs were observed for 80–120 minutes.[4]

In the experimental setting the communication with "other 13–14 year-old students" is only invoked,[5] but it does not actually take place. Nevertheless, this invoked communication structures the student pairs' activity, and more particularly it solicits a verbal formulation of the counting procedure. This is something that students do not normally do straightaway even if they are technically capable of it. At the same time, the desire to supply the "other 13–14 year-old students" with a reliable technique is likely to make the student pairs pay more attention to its formulation.

Lastly, the two students have access to as much paper as they want, but, on the other hand, to only one pencil. This constraint reinforces the cooperative nature of the situation, while at the same time giving us more direct access to the dynamic of the two confronted knowledge systems, especially in cases of decision making.

The observer intervenes only after the students have claimed that they have produced a final solution. At this stage he abandons his stance of neutrality and asks the students to deal with counterexamples that he offers. Thus there are *two different phases* during the observation, one quasi-independent of the observer and the other with a strong observer-student interaction.

The Field of Knowledge

The problem I selected calls on students' knowledge that is cultural rather than scholastic, insofar as polygons may have been studied in the primary school in the context of geometrical classification, but are no longer, for these students, a part of the taught curriculum. Sometimes they are mentioned before passing rapidly on to the study of triangles and quadrilaterals. Diagonals, however, figure in the curriculum, as they play an essential role in the study of the parallelogram.

As in the case of the solids in the 18th century,[6] students' conceptions of the mathematical content involved do not constitute a theory. In order to solve the problem, students may have to specify the objects to which it is related. In other words, to develop definitions for polygon and diagonal is part of their task.

This experimental study creates a context which is favorable for the emergence of processes like those described by Lakatos, and thus for the examination of the relevance of his approach for an analysis of students' behaviors when confronted with counterexamples to a conjecture they have produced.

TABLE I

	Types of proof [8]				
Pairs	Naïve empiricism	Crucial experiment	Generic example	Thought experiment	Proof by «reasons»
Lio-Lau	f(n)=n				
Pie-Phi	f(n)=n.s(n)				
Naï-Val	f(n) =n/2				
Pie-Mat	f(n)=n				
Nad-Eli		f(n) function of f(n-1), P_{10}			
Géo-Oli	f(n)=2f(n-1)	Oli P_{14}	(n-3)+. . . , on P_6		
Mar-Lau		Lau P_{10}	(n-3)+. . . , on P_6		
Eve-Chri	f(n)=n/2 -->			f(n)=n/2	
Bla-Isa			f(n)=n-3, sur P_7	f(n)=n/2 or n-1/2(*)	
Lyd-Mar	Lyd : f(n)=n/2		no formula for concave ones		
Ber-chr	Chr : f(n)=2n		Chr P_8	Ber : f(n)=n(n-3)/2------>	
Oli-Sté		Sté P_8	(n-3)+. . . , on P_8	s(n)=n-3 ---------------->	
Ant-Dam					f(n)=n/2
Ham-Fab					$f(n)=n^2$

(*) using a generic example, but just as a linguistic means.

2.2 *Students' Solutions and their Foundation*

The observed problem-solving procedures are closely linked to the meaning students give to the «objects» *polygon* and *diagonal*. In particular, the interpretation *regular polygon* for 'polygon', and *diameter* for 'diagonal' leads students to the conjecture f (n) = n/2.[7] Three types of solution have been proposed which «effectively» provide the required number of diagonals. They are:

$$f(n) = f(n-1) + (n-2)$$
$$f(n) = (n-3) + (n-3) + (n-4) + \ldots + 2 + 1$$
$$f(n) = (n \cdot (n-3))/2$$

The rational bases for these conjectures are more often than not empirical. But even so, the nature of the underlying proving process can vary considerably from one solution to another (cf. Table I).

Commentary

– In the column "crucial experiment" I have indicated the polygon on which this experiment is performed; in the column "generic example" I have indicated that polygon chosen by the students as being generic;
– This table does not take account of any chronology, except when an arrow expresses a change in the level of validation for a given conjecture.

Naive empiricism and generic example are dominant. The analysis of the students' dialogue shows that the lack of an operative linguistic means is one of the major reasons for the absence of proofs at a higher level. For example, use of a generic example indicates the willingness of the students to establish their solution in all its generality, but this willingness is hampered by the absence of an efficient linguistic tool to express the objects involved in the problem-solving process and their relationships.

I must emphasize that this complexity is not only linguistic but has also cognitive origins, that is the complexity of the recognition and the elicitation of the concepts needed for the proof. Linguistic constructions and cognitive constructions are dialectically related during the problem-solving process. Let us take the example of

$$f(n) = (n-3) + (n-3) + (n-4) + \ldots + 2 + 1.$$

In this case, students have to express an iterative process and to control it (number of steps, ending of the computation), but at their level of schooling they do not have the conceptual tools needed. To them, the use of a generic example seems the best means to «show» the computation procedure and to justify it.

Some of the students mention the need for a mathematical proof, but they do not try to produce one. In fact they stay at a level of proving which is consistent, on the one hand, with their level of uncertainty and what they think is needed by the situational context and, on the other hand, with the cognitive and linguistic constructions they are able to perform (Fishbein, 1982; Balacheff, 1987b).

The social interaction played a fundamental role. In few cases was it an obstacle to reaching a higher level of proof. Especially when the conflicts between two students were too strong and «too social», all means seem to be good to them as an argument, even naive empiricism.[9] But in most cases this social interaction has been the real «engine» which leads the students to an awareness of the need for proofs, forcing them to justify themselves or to elicit the rationality of the decision to be taken.

2.3 The treatment of counterexamples

Most types of treatment of a refutation as described by Lakatos, have been observed. One of the main differences is that in some cases we have to make a distinction between the treatments which follow an analysis on the part of the problem-solver, and those which are merely *ad hoc* treatment of the conjecture in order to save it in any case.

TABLE II Treatment of counterexamples, phase I

	Ant Dam	Ber Chr	Bla Isa	Eve Chr	Geo Oli	Ham Fab	Lyd Mar	Nad Eli	Oli Sté	Pie Mat	**T**	**n**
Rejection of conjecture		1	2		3	1	1		1	2	**11**	**7**
Modif. of conjecture		3	1		1	1				2	**8**	**5**
Exception					1						**1**	**1**
Condition				1			1	1	1		**4**	**4**
The definition revisited	1		2							1	**4**	**3**
Rejection of conterexample	1			1			2	1		2	**7**	**5**

T: Number of appearances of a type of treatment n: Number of pairs

Tables II and III give an account of the frequency of each type of treatment in each of the two phases of the experiment. In Table I only 10 student pairs appear out of the 14. The reasons why 4 of them are not mentioned are the following:

– Two pairs proposed at once a correct solution with reference to their conception of polygon and diagonal. Their conviction kept them beyond questions (Martine and Laura, Lionel and Laurent).

– One pair did not take any decision about a refutation they met; they asked for advice from the observer (Naïma and Valérie).

– One pair did not identify any counterexample during their problem-solving process.

These tables show important differences between "phase I" and "phase II", but these differences are not the only ones one can possibly foresee *a priori*.

It is in phase I that rejection of the conjecture is dominant, and not in phase II in which one might have expected that a counterexample proposed by the observer would provoke a rejection of the conjecture. On the other hand, the rejection of the counterexample is dominant in the phase II as well as its being treated as an exception. Two explanations can be proposed for this phenomenon:

– The rejection of the conjecture appears in the case of very fragile ones which are verified in only one case: f(n) = n (verified by P_5), f(n) = 2 • n (verified by P_7), f(n) = n2 (verified by no polygon, but conjectured by Hamdi and Fabrice ...).

– The rejection of the counterexample happens when it is opposed by a strong conjecture whether or not it is strongly established (with respect to the conceptions of the students) or correct.

TABLE III Treatment of counterexamples, phase II

	Ant Dam	Ber Chr	Bla Isa	Eve Chr	Geo Oli	Ham Fab	Lyd Mar	Nad Eli	Oli Sté	Pie Mat	Pie Phi	Naï Val	Mar Lau	Lio Lau	T	n
Rejection of conjecture		-	-		-	1	-		-	2				2	**5**	**3**
Modif. of conjecture		-	-	1	-	2				-			1	2	**6**	**4**
Exception			1		2	2							1		**6**	**4**
Condition			1	1		1	(1)	-	-			3		1	**7**	**5**
The definition revisited	1		-	3					1	-		3			**8**	**4**
Rejection of counterexample	-	2		1		1	(1)	-	4	-	2	-	1		**11**	**6**

T: Number of appearances of a type of treatment n: Number of pairs

Commentary:

In the case of Lydie and Marie, the information given is in brackets because these two students did not succeed in reaching agreement about how to treat the counterexample proposed by the observer. Lydie wanted to introduce a condition while Marie wanted to reject the conjecture and then to search for a new solution.

In the following I will present some details of the categories of treatment of a refutation. Samples of the students' behavior will be given to help the reader make sense of them.

2.4 Rejection of the Conjecture

The analysis of this type of response to a counterexample involves two different behaviors:

– The immediate rejection of the conjecture, as soon as the counterexample has been produced. This decision is more often than not taken following a naive empiricism. This behavior is coherent insofar as the observation of a few polygons was sufficient to construct the conjecture and one counterexample is sufficient to dismiss it.

Christophe defends the conjecture $f(n) = 2n$, deduced from the observation of P_7. Bertrand refutes it by P_8. In the conflict between the two students, this refutation is considered by Christophe as a crucial experiment; he then rejects his conjecture.

Lionel and Laurent defend the conjecture $f(n) = n$, deduced from observation of P_5. When the observer proposes P_6 they surrender: "*So, we were wrong*" (Lau. 204);[10] they then proceed towards a new solution.

Hamdi and Fabrice defend the conjecture $f(n) = n^2$, not from the observation of some polygons, but directly from their conception of what a diagonal consists of. When the observer proposes the counterexample P_5, they reject their conjecture and start their search for a solution again.

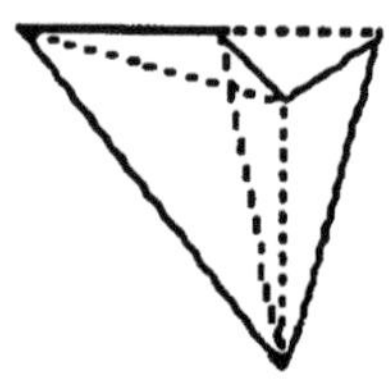

Figure 2.

– The rejection of the conjecture after an analysis of the possible origin of its refutation. This analysis provides students with new elements to restart the search for a solution to the problem:

Blandine and Isabelle defend the conjecture $f(n) = n/2$; its refutation by P_9 leads them to an analysis which reveals that when they draw the diagonals joining the vertices two by two one vertex is left. They then modify their conception of diagonal and give up their conjecture.

Georges and Olivier defend the conjecture[11] $f(n) = n - s(n)$. Its refutation by P_7 leads them to its modification into $f(n) = n/2 \cdot s(n)$, but they give it up because it appears to be impossible to divide 7 by 2. Afterwards they keep as a guide line for the search for a new conjecture, the fact that no diagonal should be counted twice.

Lionel and Laurent, whose conjecture f(n) = n • (n – 3)/2 is refuted by the observer by means of a polygon P_5 with three aligned vertices (fig. 2), keep their initial conjecture for the convex polygons and propose f(n) = (n – 3)/2) – 1 for the others. Exploring this new conjecture, they notice that it is no longer correct for any non-convex polygons, in fact "*it depends on the shape*" (Lau 970). Then they give up this new conjecture.

2.5 *Modification of the Conjecture*

The modification of the conjecture when faced with a refutation appears to be equally possible in both the phases of the experiment.[12] Three main types of such a treatment can be distinguished:

– The *ad hoc* modification which consists in a direct adaptation of the conjecture to superficial features, observed by the students, of the relations between the result expected and the one indicated by the counterexample. The origin of the new conjecture is related to the observation of one case only, thus it has a naive empiricist foundation:

Lionel and Laurent conjecture f(n) = n • (n – 3) as a solution. It is refuted by P_5; they then conclude that the solution should be f(n) = (n • (n – 3))/2 and they validate it by means of a crucial experiment on P_7. No analysis is done about the possible reasons for this division by 2, it comes from the relation they establish between the result of the counting on P_5 and what they expected. Their dialogue shows this:

"*Why would you like to divide by 2? They will not understand.*" (Lio 396)

"*We must divide by two because ... with 4, it works and with 5 also, and with 6 also ...*" (Lau 399)

– The modification which follows an analysis of the sources of the refutation with reference to the foundation of the conjecture:

Bertrand and Christophe conjecture f(n) = n • (n – 3); they meet the counterexample P_7. Then, trying to find the reasons why this polygon does not fit the formula, Bertrand discovers that "*the diagonal AF, is also the diagonal FA*" (Ber 287). Then they propose to modify the formula into f(n) = (n • (n – 3))/2, because "*otherwise some diagonal would be counted twice*" (Ber 384), thus "*that means to divide by two*" (Chr 403)

Georges and Olivier conjecture f(n) = n • s(n) after having observed that the number of diagonals is the same at each vertex of a polygon. Their analysis after a refutation by P_5 leads them towards the conclusion that "*they* [the diagonals] *appear again after ...*" (Geo 198), and they consider the new conjecture f(n) = [n/2] • s(n).

– The modifications which consist of reducing the domain of validity of the initial conjecture to a set of objects which excludes the counterexamples, and in constructing a solution specific to the objects the class of which the counterexample is considered a good representative:

Blandine and Isabelle conjecture f(n) = n/2; after its refutation by polygons having an odd number of vertices they search for a solution for this category of objects. They finally conjecture f(n) = n/2 for «even polygons» and f(n) = (n – 1)/2 for «odd polygons».

Evelyne and Christine conjecture f(n) = n/2, paying no attention to polygons with an odd number of vertices. After the refutation by the observer, who produces P_5, they reduce their solution to the polygons with an even number of vertices. For the odd polygons they propose f(n) = n ...

Nadine and Elisabeth conjecture, following a crucial experiment, that f(n + 1) = f(n) + a(n + 1).[13] The observer proposes as a counterexample a polygon P_5 with three vertices which are aligned (Figure 2); they then try to find a solution for this kind of polygon: "*The one which has a vertex inside*" (Nad 175). But they find that the solution they envisage is strongly related to the shape of the polygon, finally they surrender.

2.6 The Counterexample Considered as an Exception

This type of treatment of a counterexample is quite seldom. It appears just once in the first phase of the experiment. A possible explanation could be the belief that a mathematical assertion must not suffer any exception.

When faced with a triangle produced by the observer as a counterexample to their conjecture, **Blandine** and **Isabelle** decide that it should be considered an exception because "*it has no diagonals*" (Isa 184).

Georges and Olivier take the same decision, deciding that their conjecture is valid "*from 4 ... because triangles are particular cases*" (Geo 508). When faced with a polygon P_5, three vertices of which are aligned, they state that "*it is again a particular case*" (Oli 590), but "*that's a lot of exceptions*" (Oli 625). They are not really very confident: "*Perhaps it is the good solution, even if there are a lot of exceptions*" (Oli 645). Then they try to find a specific solution for these cases. Failing to do so, they finally keep on with the decision to consider these counterexamples as exceptions.

When **Hamdi** and **Fabrice** conjecture $f(n) = n^2$ as a solution; they envisage the case of the triangle but "*any polygon should not have zero diagonals*" (Ham 222). When the observer proposes the case of P_4 they claim again that "*it is a particular case*" (Fab 279), and they confirm their initial solution but "*except the particular cases ... triangles and equilaterals*" (Ham 389). They continue to consider these polygons as exceptions, even when they propose the solution f(n) = (n • (n – 3))/2 for which it is no longer the case.

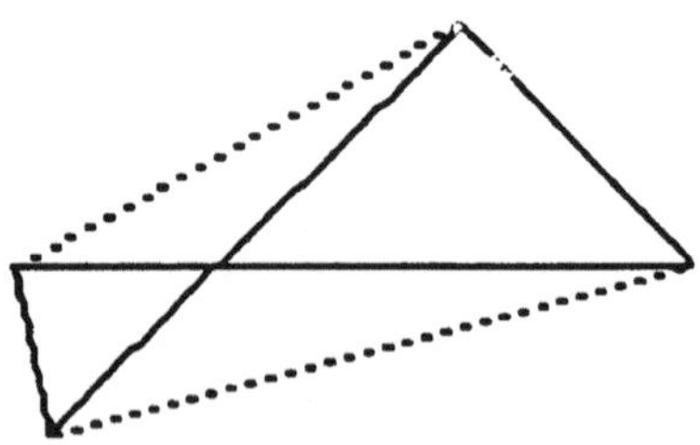

2.7 Introduction of a Condition

The decision to introduce a condition in the statement of the conjecture follows two types of strategies:

– Some students try to find a condition by means of an analysis of the counterexample to bear down the class of objects of which it is a representative:

Hamdi and Fabrice introduce a condition in order to brush aside the objects represented by the counterexample the observer proposes to them (fig. 3): "*Except these which have a point in their middle*" (Ham 616). Such a condition has nothing to do with what could constitute the foundation of their conjecture.

Naïma and Valérie protect their conjecture against the counterexamples produced by the observer, adding *ad hoc* conditions: "*all the edges should be of the same size*", "*the edges should be on a circle*", "*to a vertex belongs only one line*". This strategy is absolutely resolute "*I don't mind, I answer with conditions*" (Val 504).

– Some students try to find a condition by means of an analysis of the refuted conjecture and its foundations:

Blandine and Isabelle decide to introduce a condition against non-convex polygons, one of them having been produced by the observer as a counterexample, because "*the diagonals should be inside, as they should intersect just once*" (Isa 206). They then decide that the polygons to which the conjecture apply must be convex.

Olivier and Stéphane conjecture that the larger the number of vertices of a polygon, the larger is the number of its diagonals. That happens to be refuted by a polygon P_7 which has some vertices on the same line. The students analyze the origin of this refutation and introduce a condition: "*We have to say that it works if the points are not on the same line*" (Sté 77).

The decision to introduce a condition seemed "not allowed" to some students, because of the experimental contract.[14] For example, the conjecture $f(n) = n/2$ having been refuted, Evelyne proposes to introduce a condition on the number of vertices of a polygon: "*It is always divided by two ... so the number should be even*" (Eve 85). But the students finally do not introduce this condition, because no restrictions on the type of polygon have been stated in the text of the problem: "*It might not be so, because there* [in the text of the problem] *it is not said that the polygon has ... that the number of vertices is even*" (Eve 140).

2.8 The Definition Revisited

Most of the students posing the problem of what a polygon is, or trying to elicit a definition of polygon or diagonal, are those whose conjecture is $f(n) = n/2$ (6 cases of the 9 observed). Their uncertainty about what a polygon is is noticeable from the very beginning of the experiment, but it is only after a refutation that some of them clearly state this problem of definition:

The doubt about what a polygon could be urges **Antoine** and **Damien** to pose the problem of its definition: "*It could take a long time if we search like that ... if we have not the exact data*" (Dam 77). Taking P_8 as a prototype, they put aside the odd polygons: "*It cannot be an odd number as far as we say that the edges should be parallel two by two*" (Dam 221).

Actually this problem turns into being part of the «game»: "*the rule tells that we are the teacher .... We tell him* [the observer] *a polygon is what we have given, and he should find whether our definition is correct or not*" (Ant 383).

Blandine and **Isabelle** straight away pose the problem of the definition of polygon and diagonal. A first counterexample they encounter against their conjecture ($f(n) = n/2$) leads them to the following statement: "*it* [a diagonal] *joins two vertices*" (Bla 18). Later a new conjecture having been formulated ($f(n) = n \cdot (n-3)$) and then refuted by P_7, they refine their definition: "*to each vertex belongs one diagonal*" (Isa 70). Relying upon that definition, they return to their initial conjecture which they now hold quite strongly.

Naïma and **Valérie** conjecture $f(n) = n/2$, after the refutation by the observer they state that "*it is perhaps necessary to know the definition of polygon*" (Val 285). In fact, they come to the idea that the problem is basically a problem of definition and, after having stated a definition, they modify it to cope with the counterexamples produced by the observer. But they recognize that the definition they propose is "*what we think*" (Nai 444). This focus on the problem of definition could be related to the wish "*to find something valid for all polygons.*"

2.9 Rejection of the Counterexample

This way of treating a counterexample is the most widely distributed among the observations I have made. It appears at least once in each of 11 pairs out of the 14 observed. On the other hand, its presence is more frequent in phase II of my experiment during which the students were perhaps more eager to defend their conjecture against refutation by the observer. But not all these reactions should be considered the same.

Three main categories can be differentiated:

– The counterexample is rejected after an analysis which reveals that actually it does not refute the conjecture: It is shown that the refutation is made on the ground of a mis-interpretation. The possibility of such an analysis was opened by the case of a polygon P_5 with three aligned vertices, proposed by the observer:[15]

Bertrand and Christophe discover the hidden diagonal in the drawing by the observer of a polygon P_5 with three aligned vertices. They suggest that it might be good to inform the recipients of their message of such a possible event. When faced with a new counterexample produced by the observer, a polygon P_4 with edges crossing elsewhere than the vertices, they make an attempt to show the same misinterpretation on the part of the observer. They fail to do so ...

– The counterexample is rejected with reference to a precise conception of what is a polygon: Conception which may have been formulated previously as a definition:

Pierre and Mathieu, who have conjectured $f(n) = n/2$, reject a convex polygon P_5, because it appears not to be regular: "*The edges are not exactly equal*" (Mat 63). About the same polygon Mathieu even suggests that "*maybe a polygon with 5 edges doesn't exist*" (Mat 135). Actually the difficulties with this counterexample will be overcome later while revisiting the definition of diagonal.

Olivier and Stéphane reject the counterexamples proposed by the observer, analyzing them as joint figures but not real polygons: "*That's not a polygon*" (Oli 768) "*Wait ... a polygon is a thing ... here it's not a polygon ... they are two triangles ...* [to the observer] *Is that a polygon?*" (Sté 769) Actually, in order to reject the counterexample the students have to raise the problem of the definition of a polygon: "*A point must be related to no more than two other points*" (Sté 382)

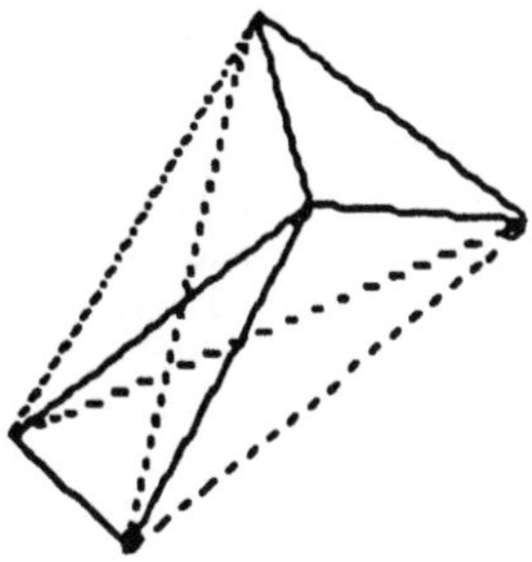

Figure 4.

– The last category is the one Lakatos named «monster-barring», the counterexample is in this case rejected by students without any further considerations about the definition or any debate about their conceptions. It is, in particular, the case for the triangle because of its lack of diagonals.

Evelyne and Christine reject the triangle as a counterexample to their conjecture f(n) = n/2 because "*the triangle has no diagonal*" (Chr 113). Later on, the triangle being met again, they reject it definitely and explicitly: "*the triangle is not a polygon*" (Chr 473).

Hamdi and Fabrice reject the triangle despite the fact that the observer has shown that it fits their latest conjecture. Since it was a real counterexample to a previous conjecture, they do not want to consider it again and "*whatever it is, in a triangle there are no diagonals ... so we are not interested*" (Fab 553).

III. STRATEGIES IN THE TREATMENT OF A COUNTEREXAMPLE: STABILITY AND DISPERSION

When the number of counterexamples met by a pair of students is large, it is of interest of examine their global strategy and the distribution of the type of treatments they have chosen. With respect to this question two types of categories should be recognized:

– Stability of treatment: Few different types of treatment are used by the students. The origin of this stability, which has been observed in three pairs, is quite different from one case to another: In the case of Bertrand and Christophe, or Olivier and Stéphane, it is due to the success of the first decision they took about a counterexample; in the case of Naïma and Valérie it is a consequence of their commitment in a «definition game».[16]

– The diversity in the treatment of the counterexamples cannot be characterized in a few words; I give an account of the pairs which appear to be the most significant with respect to this kind of behavior.

Evelyne and Christine: These students construct and defend the conjecture f(n) = n/2, first on the basis of naive empiricism, and then on that of a thought experiment related to the conception of diagonal which they elicit after the refutation of their conjecture by P_7: "*A diagonal is a straight line which belongs to a vertex of a polygon and which cuts its surface in two pieces.*"

But they do not share the same point of view: One of them would like to introduce a condition in the statement of the conjecture, the other would prefer to consider the definition again. Finally, it is this last decision which is taken, following an argument of Evelyne, who says that if a condition was necessary, it would have been stated in the description of the task. In other words, the students have not the right to restrict the scope of the polygons which fit the conjecture, they have to give a solution for all of them. Actually, the definition they choose restricts the set of the polygons to those for which the conjecture is valid: "*If we say that a polygon is ... a thing whose sides are always parallel two by two ... then it is not necessary to be more precise. We have just to divide by two*" (Chr 201).

But on the other hand, the case of the triangle is not dealt with referring to the definition. It is rejected as a polygon because it has no diagonals, it appears to the eyes of the students as a kind of monster.

Later on a counterexample, P_5, because it is produced by the observer, imposes itself as a polygon. Then the initial definition is rejected, and a new one is considered: "*for sure a polygon can have any number of diagonals ... but it should be regular*" (Chr 270). Thus the only solution to save the conjecture is to introduce a condition and to search for a solution specific to odd polygons.

This last solution starts from a student's conception of diagonal as an «axis of symmetry». In case of P_5 it corresponds to a drawing they make (fig. 5). The solution they conjecture is f(n) = n for odd polygons, first proved by means of a thought experiment and afterwards proved «by reasons».[17] The uncertainty which remains is an uncertainty about the premisses of the students' reasoning: Definition of polygon and of diagonal. It is on the basis of these definitions that they then treat the counterexamples produced by the observer. These definitions are even written into their message: "*If the number of vertices of the polygon is even: You divide the number by two and you will obtain the number of its diagonals. If the number of vertices of the polygon is odd: The number of the diagonals is equal to the number of the vertices. A polygon is a geometrical figure which can have any number of vertices but whose edges must be equal*"

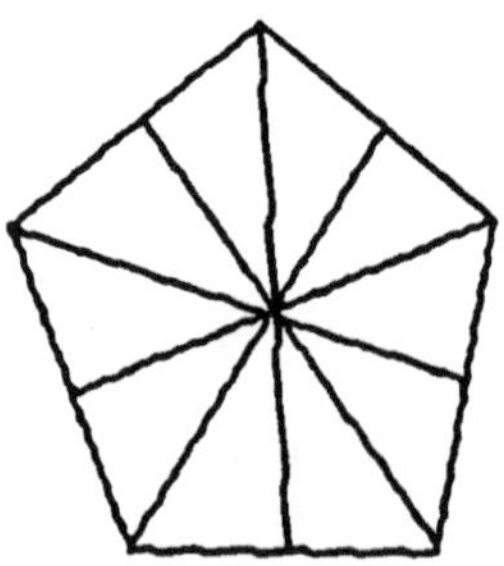

Figure 5.

For Evelyne and Christine, dealing with a counterexample is directed by what I would like to call a *problématique*[18] of the definition which originates in their reading of the experimental contract, following which they think that they cannot introduce any reduction of the field of polygons to be considered. The only one which escapes this treatment is the triangle which appears to them as a "monster" and which is definitely rejected as such.[19]

Hamdi and Fabrice: From their conceptions of polygon and diagonal, these students conjecture: $f(n) = n^2$. After a refutation they encounter with P_4, they first try to modify this conjecture. Failing in doing so, they decide to submit it to the observer. His refutation by P_5 as a counterexample leads them to give up this first conjecture. The two refutations lead them to the problem of the definition "*he* [the observer] *should have given the definition of polygon*" (Ham 228).

They then search for a new solution taking as (implicit) basis what appears to be a correct conception of a convex polygon. Their new conjecture is $f(n) = n \cdot (n-3)$, which is refuted by the observer by the counterexample P_6. They modify their conjecture into $f(n) = (n \cdot (n-3))/2$. Actually, it is an *ad hoc* modification which is elaborated straightaway from the arithmetical relation they establish between the result obtained (that is, 9) and the one they had computed (that is, 18). This conjecture is then confirmed in the case of P_5. Then they envisage a crucial experiment on P_{16} which they finally abandon. They state in their message that the triangles and the quadrilaterals are outside the domain of validity of their conjecture. They change their mind in the case of the quadrilateral after an example produced by the observer, but they stand on their position in the case of the triangle.

The other counterexamples produced by the observer are the following drawings (Figure 6).

Figure 6.

First, on the one hand, Fabrice refuses to accept them as being polygons, but on the other hand, Hamdi treats them as exceptions. Then Fabrice remembers that there are polygons with edges intersecting elsewhere than vertices, so he revises his point of view and he considers further the drawings of the observer as being polygons. Thus the task for the students becomes that of stating a condition to push aside this kind of object: "*it does not work for polygons which have a point in the middle*" (Fab 682). We should notice that this condition is in no way related to an analysis of the conjecture and its possible foundations, but consists in an *ad hoc* adaptation to the apparent features of the counterexamples.

The rejection of their initial conjecture by Hamdi and Fabrice is due to their uncertainty about their conceptions which underlie it. It is this same uncertainty which leads them to raise the problem of the definition. The modification of their second conjecture f(n) = n • (n – 3) appears to be unavoidable as the counterexample P_6 cannot be rejected as not being a polygon, but this modification is *ad hoc* in order to save the conjecture. On the other hand the counterexamples produced afterwards by the observer do not lead to new modifications, they are rejected: The conditions introduced by the students being nothing but *ad hoc* means to preserve the conjecture.

Then, the treatment of the refutations is dominated by the students' analysis of the objects in question, perhaps to the detriment of an analysis of the rationality of the conjecture.

Pierre and Mathieu: These students do not share the same conception of polygon and diagonal; this disagreement is at the core of their conflict. According to Pierre, a polygon is a «geometrical figure with several sides» but diagonal is taken as a synonym of oblique (actually, a line non-parallel to the edges of the polygon). For Mathieu, a polygon is nothing but a regular polygon whose diagonals are diameters. For a while Mathieu considers the diagonals as being axes of symmetry; this conception is rejected by Pierre because of his own conception. Finally, this conception is rejected after confrontation with the case of the square.

The first conjecture of these students: f(n) = n/2, is refuted by P_5. The treatment of this counterexample is at stake in a debate between Pierre and Mathieu about what is or is not a diagonal, a polygon, or about the relevance of a proposed response:

– Mathieu proposes to reject the conjecture, but Pierre opposes that: It is so convincing that that should not be done;
– Pierre suggests treating it as an exception but Mathieu refuses because a mathematical assertion must apply to each of the mathematical objects concerned;
– Mathieu proposes to reject it as a non-polygon, but Pierre refuses, referring to the etymology of the word «polygon»
– Pierre proposes to stretch the concept of diagonal (including "*half-diagonals*"), but such an extension clearly appears to be too *ad hoc*.

Finally they search for a specific solution for the odd polygons, despite the fact that they consider that a «mathematical» solution should be expressed by a unique formula. They conjecture f(n) = n/2 for even polygons, and f(n) = n – 1 for odd polygons (which is in fact an *ad hoc* adaptation of the initial conjecture). But Pierre, challenged by the failure of his attempt to stretch the concept of diagonal, finally changes his mind about what a diagonal consists of, accepting the proposal of Mathieu. They propose to the observer: f(n) = n/2 for even polygons, and f(n) = n for odd polygons (which is again an *ad hoc* adaptation to the case of P_5).

After the next refutation of that conjecture by the observer, Pierre surrenders, while Mathieu tries to find **one** solution. Finally they abandon it.

The conflict between Pierre and Mathieu is, first of all, a conflict between two different conceptions about what a polygon or a diagonal is. This conflict leads them to debate the problem of the definition of these objects. This problem being solved, and the two students' agreement about what is allowed in mathematics being found (status of the exceptions, unique formula), overcoming of a refutation consists in the search for specific solutions, and then for **one** solution. For Pierre, the fact that each of their solutions is refuted leads to the idea that the situation is nothing but a game in which they have to "guess" a solution. Actually, he abandons the search to try to obtain **the** solution from the observer.

IV. CONCLUSION

Three factors appear to determine students' choice in their treatment of a refutation:

– **The analysis with reference to the problem itself.** It gives a central place to the discussion on the nature of the objects concerned, and thus on their definition. This analysis potentially leads to any type of treatment which could be considered, none of them being privileged *a priori*. The choice the students make can be understood only in the light of a local analysis or of the specific character and knowledge of each of the individuals. The type of treatment can change in the course of the problem-solving process: Modification of the definition followed by the introduction of a condition or a modification of the initial conjecture when the conceptions have been stabilized. The origins of the choice between the introduction of a condition and the search for a specific solution, and the modification of the conjecture, cannot be traced in the data gathered. What could be conjectured is that when the refutation might brush aside a wide range of polygons (with respect to the students' conceptions), then a modification of the conjecture is decided which consists in a specific solution (extension to the odd polygons in case of $f(n) = n/2$, search for a solution for non-convex polygons).

– **The analysis with reference to a global conception of what mathematics consists of.** This could be a serious obstacle to some of the treatments of a refutation: Refusal to treat the counterexample as an exception, refusal of a solution which cannot be expressed by a unique formula, etc.

– **The analysis with reference to the situation**. What is in question is mainly the didactical contract which leads students to favor some treatments of the counterexample (definition game, riddle game in which they abandon their solution quite easily) or constitutes an obstacle to others (refusals to introduce a condition because it has not been stated in the problem statement). Whatever is the case, the role played by the experimental contract should not be over-estimated. In particular, it should be noticed that the fact that treating the counterexample as an exception or its rejection, during phase II of the experiment, confirms the result already known about the fact that one counterexample is generally not sufficient for students to call a conjecture into question (Burk 1984, Galbraith 1979). The fact that this treatment has been observed in other, and quite different, experimental settings, suggests that it is not specific to the experimental contract in this situation as it might be conjectured at first.

One of the questions I have examined is that of a possible influence of the type of conjecture on the choice of a treatment of a refutation. One hypothesis might be that if the conjecture is false, then its rejection or its modification or a revision of the definition should be dominant, but that, if the conjecture is correct, then the rejection of the counterexample should be dominant.

Actually, I have observed that conjectures like $f(n) = n$ or $f(n) = 2n$ are abandoned after their refutation. In fact, such conjectures are very fragile in so far as they are verified by only one polygon. It is quite different when the conjecture is verified by a large set of polygons, like $f(n) = n/2$. The strength of that conjecture comes from the fact that it is related (explicitly or not) to the conception of a polygon as a regular polygon and the diagonal as a diameter.[20] On the other hand, I have checked the type of treatment of the counterexample against the type of foundation of the conjecture; it appears that no treatment of the counterexample is privileged, whether the foundation of the conjecture is naive empiricism or a thought experiment.

I have also examined the case of the correct conjectures. It should be noticed that they implicitly refer to a correct conception of polygon and diagonal. These conjectures are the result of a process which has occupied all the first phase of the experiment, whether they have been constructed deductively or are the result of a dialectic between successive attempts and their refutation. The treatment appears to be far less varied than in the case of the conjecture $f(n) = n/2$. A first type of treatment which is dominant is the rejection of the counterexample after its analysis relative to the students' conceptions, a second type of treatment is to consider the counterexample as an exception or to introduce a condition (actually this last treatment could be a way for some students to escape particular cases).

Whatever is the level of proving involved in the problem-solving process, it appears that the robustness of the students' conceptions and the existence of a large domain of validity of their conjecture lead to privilege the treatments which consist of keeping aside counterexamples which appear to be some kind of «monsters»[21] in the eyes of the students. On the other hand, I have found that in the case of the conjecture $f(n) = n/2$, the limitation to even polygons is associated with an important uncertainty about the definition, which is often reconsidered after the refutation.

Mathematicians, Lakatos considers, share almost the same rational background. In the case of the students the background is not the same: Naive empiricism, or pragmatic validations, can be the basis of their proof and can constitute the roots of their belief in the truth of a statement. If this is so, the discussion of the criticism of a counterexample or of the modification of conjecture could appear less relevant to the teacher than the discussion of its «logical» background. How can we escape the fact that faced with a counterexample produced by the teacher, students claim that it is a particular case when, in fact, what should be questioned is the naive empiricism on which their conjecture is based? At a higher level of schooling this problem can still appear: A student may be discussing the legitimacy of a counterexample, when it is his or her understanding of the related mathematical knowledge that should be questioned.

To base the learning of mathematics on the students' becoming aware of a contradiction requires that we take into account the uncertainty about the ways they might find to overcome the contradiction. If, as I believe, we cannot assume that there is strict determinism in cognitive development, what could be the role played by a particular situation? The teacher's interventions will be fundamental. The way he or she manages the teaching situation may bring the students to see that their knowledge and the rationality of their conjectures must be questioned and perhaps modified, because no *ad hoc* adaptation of a particular solution, or its radical rejection, can by itself lead to a conceptual advance.

NOTES

* *Acknowledgement*: I am greatly indebted to David Pimm and Ernst Von Glasersfeld for helping me with the English version of my text.

1. Quoted by Hadamard (1959, p. 34).
2. By "proof" we mean a discourse whose aim is to establish the truth of a conjecture (in French: *Preuve*), not necessarily a mathematical proof (in French: *Démonstration*).
3. Indeed such a schema just evokes the range of possible consequences, but it is sufficient to give an idea of what we call the openness of the possible ways to treat a refutation.
4. Cf. (Balacheff, 1988) for a complete report on this research.
5. That is to say: Communication is fiction, the other participants (the receivers) do not exist. (Mugny 1985, chap. 10). We call such a situation: *Situation for Communication.*
6. In 1758 Euler published his *Elementa doctrinae solidorum* which was about "corpora hedris planis inclusa". It is there that we find the first formulation of the famous theorem.
7. "f" is the name we will use for the function which relates the number n of vertices of a polygon to the number of its diagonals.
8. P_n stands for "a polygon with n vertices".

   By *naive empiricism* we mean the situation in which the problem-solver draws the conjecture from the observation of a small number of cases (for example that $f(n) = n$ because P_5 has 5 diagonals). We use the expression *crucial experiment* when the problem-solver verifies the conjecture on an instance which "doesn't come for free", here the problem of generality is explicitly posed. The *generic example* involves making explicit the reasons for the truth of the conjecture by means of actions on an object which is not there in its own right, but as a characteristic representative of its class. The *thought experiment* invokes action by internalizing it and detaching itself from a particular representation. (For a precise definition of the different types of proofs, see: Balacheff, 1987a and 1988.)
9. See, footnote 8.
10. "Lau 204" is a reference to the protocol of this pair of pupils, it means "Laurent is speaking, action No. 204".
11. Here "s" is the name of the function which associates with the number n of the vertices of a polygon, the number s(n) of its diagonals at one vertex. That means that pupils have recognized the invariance of the number of diagonals belonging to each vertex of the polygon.
12. cf. Table II and Table III.
13. Here "a" is the name of a function which gives the value of the difference between the number of diagonals of P_n and that of P_{n+1}.

14. The relationships between the actors involved in the situation (the two pupils and the observer) are determined by rules which are both explicit (like what is said about the task), and implicit (like those suggested by the aim of the experiment). All these rules constitute what we call the «experimental contract» (Balacheff and Laborde, 1985).
15. cf. Figure 2.
16. Against each counterexample proposed by the observer, they enrich their definition with a new *ad hoc* clause.
17. That is what we have called in French: *Preuve en raison*, that is a proof by what could be seen as a real computation on statements (in French: *Calcul sur des énoncés*).
18. By *«problématique»* we mean the framework within which problems are stated and recognized as being relevant.
19. We should remark that, following their definition of polygon and diagonal and the solution they proposed for odd polygons, the triangle is no longer a counterexample. On the other hand, there is a contradiction they do not examine, between what they know about the triangle and the fact that the conceptions they have elaborated lead to 3 diagonals for a [equilateral...] triangle.
20. Such a conception is reinforced by the prototype of the square.
21. That is, a «monster» with respect to their conception of what a polygon is. For example, a polygon must have diagonals (actually the problem underlying this view is that of the interpretation of «zero»).

REFERENCES

Balacheff, N. (1987a). "Processus de preuve et situations de validation". *Educational Studies In Mathematics* 8, pp. 147–76.

— (1987b). "Cognitive Versus Situational Analysis of Problem- Solving Behaviors". *For The Learning of Mathematics*, 6, 3, pp. 10–12.

— (1988). *Une étude des processus de preuve en mathématiques chez des élèves de Collège.* Thèse d'état, Université de Grenoble 1.

Balacheff, N. and C. Laborde (1985). "Langage Symbolique et Preuves dans l'Enseignement Mathématique: Une Approche Socio-Cognitive". In Mugny G., *Psychologie Sociale du Développement Cognitif.* Berne: Ed. Peter Lang.

Burke, M.J. (1984) *The Use of Counterexample Logic By Adolescent.* Ph.D., The University of Wisconsin, Madison.

Fishbein, E. (1982), "Intuition and Proof." *For the Learning of Mathematics.* 3, 2, pp. 9–18 and 24.

Galbraith, P.L. (1979). *Pupils Proving.* Shell Centre for Mathematics Education, University of Nottingham.

Grize, J.B.; Piéraut-le Bonniec (1983). *La Contradiction.* Paris: PUF.

Hadamard, J. (1959). *Essai sur la psychologie de l'invention dans le domaine mathématique.* Paris: Gauthier-Villars, 1975.

Inhelder, B. and D. Caprona (1985). "Constructivisme et création de nouveautés". *Archives de psychologie.* 53, 204, pp. 7–18.

Lakatos, I. (1976). *Proofs and Refutations.* Cambridge: Cambridge University Press.

Mugny, G. (1985). *Psychologie Sociale du développement cognitif.* Berne: Ed. Peter Lang.

Piaget, J. (1974). "Les Relations entre Affirmations et Négations". *Recherches sur la contradiction.* Vol. 2, Paris: PUF.

— (1975). "L'équilibration des structures cognitives." *Etudes d'épistémologie cognitive.* Vol. XXXIII. Paris: PUF.

Robert, A. (1982). *L'acquisition de la notion de convergence des suites numériques dans l'enseignement supérieur.* Thèse d'état. Université de Paris VII.

LEARNING TO LISTEN: A STUDENT'S UNDERSTANDING OF POWERS OF TEN [1]

JERE CONFREY

Cornell University

I. INTRODUCTION

Basic to applying the constructivist epistemology to mathematics education are two claims:

1. When students genuinely engage in solving mathematical problems, they proceed in personally reasonable and productive ways.
2. Researchers and teachers must learn to listen and to hear the sense, and alternative meanings in these approaches.

Although necessary for the application of constructivist epistemology to research or teaching in mathematics education, acceptance of these two claims is not sufficient to ensure constructivist practices; for, they can be interpreted in multiple ways.

In this paper, I will:

1. contrast alternative interpretations of these claims as they might be understood in the traditions of discovery learning, problem solving and misconceptions with a constructive interpretation.
2. summarize some of the basic assumptions for conducting an investigation of students' conceptions within the constructivist framework, and
3. provide an illustration using interview excerpts from a student who was asked to represent on a number line a series of historical events given in scientific notation.

Examining only briefly the similarities and contrasts between constructivism and other perspectives on student learning in mathematics has certain advantages and hazards. It can assist us in learning how constructivism complements or contrasts with other widely held approaches (as a puzzle piece fits in a jigsaw puzzle) in preference to presenting, in isolation, its own tenets and claims. The contrasting theories were selected because they share many assumptions with constructivism, and yet, their treatment of the epistemological character of mathematics or of the socio-psychological genesis of ideas is different. An inherent hazard of a brief presentation is that the constrasting theories may appear oversimplified and too easily dismissed without an appreciation for their role in the progress of educational thought. To compensate for this, references to fuller treatments and reviews of the theories are provided.

E. von Glasersfeld (ed.), Radical Constructivism in Mathematics Education, 111–138.

II. DISCOVERY LEARNING

Many people will read the first claim for applying constructivisim as parallel to one of the basic assumptions behind "discovery learning" (Bruner, 1960, 1966). Discovery learning as it was advanced in the 1960's referred to "methods that permit a student to discover for himself the generalization that lies behind a particular mathematical operation" and it was contrasted to "the *method of assertion and proof* in which the generalization is first stated by the teacher, and the class is asked to proceed through the proof." (Bruner, 1960, p. 21) The methods of induction (drawing generalizations from an array of examples) and deduction (drawing conclusions from premises) formed the basis of the contrasting positions, with discovery learning relying on induction. As a proponent of discovery learning, one might interpret the first claim to mean: if a concept is inductively demonstrated through the use of mathematical problems, the students will actively participate in the process of "uncovering" that generalization in resourceful and effective ways.

Discovery learning certainly rested on some assumptions that constructivists share. It stressed the importance of: 1) involving the student actively in the learning process; 2) emphasizing the process of "coming to know" over the rapid production of correct answers; and 3) extracting and making increasingly visible the structure of a concept. Its basic claim, that students could be guided to discover rather than be told, was a radical departure from the assumption that mathematics could only be taught through direct instruction following by drill and practice.

However, the statement in the first claim, "genuinely engaged in solving a problem", entails, for the constructivist, more than reasoning inductively to conclude with a predesignated generalization carefully manufactured by using a set of examples. Advocates for discovery learning assumed that behind the set of examples lies a generalization, awaiting discovery. That generalization, logically required by the structure of the concept, was assumed to be retrievable relatively automatically and uniformly across students by their application of inductive technique. The argument was that through such a process, the student would be more likely to internalize the results. Discovery learning was a model for promoting more effective learning – the epistemological content, (the claims about the mathematical knowledge to be learned) remained relatively untouched.

Most often, the applications of discovery learning did make the generalization itself evident as a repeated pattern. But, making such generalizations still did not ensure any deep understanding, for they left the content undisturbed. In striving to have students make inductive generalizations, the educators often neglected to assist students in developing a sense of the functionality of the concept – its purpose and usefulness as an explanatory construct, its epistemic character. For example, one can conclude inductively that changing a percent to a decimal requires one to move two spaces to the left without gaining any insight into the advantage of place value (even if the process includes use of the algorithm for dividing by 100). However, if the students do not see the efficiency and elegance of the multiplicative structure of place value (as opposed to the awkwardness of the cumulative structure of systems like Roman numerals), the inductive generalization will be only minimally more meaningful to the student. In discovery learning, revisions in the mathematics itself were not considered necessary; only the logical process used to learn was altered, as induction replaced deduction.

For the mathematicians and cognitive psychologists working on the ideas of discovery learning, mathematical ideas were seen to lie within a larger conceptual field whose structure was hierarchical, consistent and logically necessary. "Big ideas" such as sets, functions, deductive systems and properties were thought to provide a powerful and parsimonious structure into which individual generalizations by student would be slotted. Thus, "discovery learning" entailed a commitment to a form of Platonic idealism. The truth value of a mathematical claim was derived from outside the framework of human experience. When students "discovered" a mathematical necessity, there was no need to consider the epistemological status of such a claim – its certainty, necessity, unformity and stability were assured by the membership in the class of mathematical truths.

III. A CONSTRUCTIVIST VIEW OF MATHEMATICS

The view of mathematics underlying discovery learning is at variance with that of the constructivist. For the constructivist, mathematical insights are always constructed by individuals and their meaning lies within the framework of that individual's experience. Students' explanations, their inventions, have legitimate epistemological content and are the primary source for investigation (other potential sources include the beliefs of teachers and mathematicians). For the constructivist, mathematical ideas are created and their status negotiated within a culture of mathematicians, of engineers, of applied mathematicians, statisticians or scientists, and, more widely, in society as a whole, as it conducts its activities of commerce, construction, and regulation.

The constructivist does not deny those who study or practice mathematics the authenticity of their experience of profound certainty when they have uncovered a mathematical claim. The sense of profound certainty, the documentation of concurrent inventions, the unpredicted convergence of apparently disparate fields, the tenacity of certain mathematical constructs all require explanation. But, equally important, the constructivist does not disregard numerous examples of mathematical claims by courageous individuals, denied at the time as impossible, which eventually led to significant advances; nor does s/he wish to ignore the evidence of views, widely held as self-evident at one time, which later were viewed as in error or outside the mainstream of important mathematical thought. (Klein, 1980; Davis and Hersh, 1981; Trudeau, 1987).

In rejecting the idea of Platonist truths whose existence is independent of humanity, the constructivist relies on explanation based in the interplay between social negotiation of meanings and individual creativity and genius. In constructivism, the development of mathematical ideas is explained through their cultural history of negotiation (Lakatos, 1976; Toulmin, 1972); particular attention is given to examining how multiple systems of representation, symbolism and tools create occasions for the convergence of meanings – a convergence that is created through the ways in which those who practice mathematics weave together notational, linguistic, manipulative and operational forms of description. Regrettably, the Platonic view of mathematical truth (as existing outside of space and time and yet accessible through the purest forms of human abstraction) is reinforced by the scarcity of historians and mathematicians capable of "rational reconstruction" (Lakatos, 1976; Unguru, 1976). The work demands individuals capable of unraveling rich and complex chains of reasoning winding over long periods of time of time (Elkana, 1974; Toulmin, 1972; Lakatos, 1976; Feyerabend, 1978; Tymoczko, 1986).

Relying on the philosophers of science to provide the historical analysis, educational constructivists engage instead on a pursuit of documenting and describing the course of development of mathematical ideas in children, adolescents, and adults. It is this "genetic epistemology" (Piaget, 1970), a description of how people come to understand the epistemic structure of mathematical or scientific ideas, to which the constructivist programme is committed. The constructivist view of development of mathematical knowledge, which is assumed to entail similar influences whether it be studied as historical (across communities over significant spans of time) or developmental (among individuals participating in society over life spans), is taken as an assumption which undergirds the constructivist programme. It can be stated as follows:

Assumption One: Constructivists view mathematics as a human creation, evolving within cultural contexts. They seek out the multiplicity of meanings, across disciplines, cultures, historical treatments, and applications. They assume that through the activities of reflection and of communication and negotiation of meaning, human beings construct mathematical concepts which allow them to structure experience and to solve problems. Thus, mathematics is assumed to include more than its definitions, theorems and proofs and its logical relationships – included in it are its forms of representation, its evolution of problems and its methods of proof and standards of evidence.

Accepting this description of mathematics has implications for the constructivist examining students who are engaged in solving problems. The constructivist does not expect a student to produce textbook generalization – in form or content. S/He begins with the assumption that what a student does is reasonable and then seeks to describe it from the student's perspective. The investigation is assumed to have epistemological content – not just psychological content – in that recognizing, unearthing and giving validity to a student's method requires one to confer on it the status of genuine knowledge. If the student fails to ever articulate it, or to continue to demonstrate its usefulness and viability in solving his or her problems, then the interviewer may abandon this assumption. Such a situation is, however, surprisingly rare. Thus, unlike in discovery learning where the generalization sought after is presumed to be predictable prior to the investigation, the constructivist is engaged in a processs of invention – invention of his/her own models for explaining students' actions and words.

Assumption Two: In examining a student's understanding of a mathematical concept, a constructivist seeks to represent how a student approaches the mathematical content. S/He expects diversity – and idiosyncratic rationality. The interviewer's knowledge of the mathematical content, complete with multiple representations, competing interpretations, various applications – guides the inquiry, but his/her intent is to examine the student's use of examples, images, language, definitions, analogies etc. to create a model which may well transform the interviewer's own understanding of the mathematical content in fundamental ways.

III. PROBLEM SOLVING

A second interpretation of the first claim (p. 1) might be that it endorses problem solving approaches (Polya, 1945, 1962; Goldin and McClintock, 1979; Silver, 1982; Schoenfeld, 1985) Problem solving advocates argue for explicit instruction in the solving of mathematical problems. Problem solving is described through its use of heuristics and metacognitive strategies frequently set within Polya's four-step process (understanding the problem, devising a plan, carrying out the plan and looking back.)

The paradigm for problem solving shares with constructivism some basic assumptions that include stressing: 1) the value of a problem as "a means of finding a way out of a difficulty, a way around an obstacle, attaining an aim that was not immediately obtainable." (Polya, 1972, p.v.); 2) the importance of problem clarification. "The worst may happen if the student embarks upon computations or constructions without having *understood* the problem". (Polya, 1945, p. 6); 3) the significance of elucidating strategies that are typically tacit but effective; and 4) the value of looking back, i.e. reflecting on one's solution path.

These shared assumptions constituted a profound shift in our perspectives on mathematics education. Identifying heuristics involved mathematicians and educators in a process of self-reflection which led to the expression of personal epistemological assumptions concerning the structure of mathematical problems. Moreover, by claiming that problem solving could be taught, problem solving theories weakened the insipid belief that mathematical ability was a holistic character trait conferred (or not) at birth. Remarkably, problem solving approaches began with the assumption that the bastion of mathematical creativity, the ability to solve problems, could be taught. Thus, the tradition of problem solving introduced talk-aloud methods of examining a person's talk during the activity of problem solving. Finally, by focusing on problem solving, advocates clearly demonstrated teachers' premature concern with proof, public standards of evidence while at the same time they ignored the process of invention – one which was to rely on intution, insight and heuristic.

Just as proponents of discovery learning assumed a specific structure inherent in a concept, problem solving advocates tended to assume a well-defined structure inherent in a problem. Heuristics, (try a simpler problem, consider special cases, use auxiliary information) supported the perspective that mathematical systems were well-organized structures in which problems were suspended. In Polya (1945), we witness this formalization of a problem in his characterization of a student's initial understanding of a problem as "incomplete". He states, "Our conception of the problem is rather incomplete when we start the work; our outlook is different when we have made some progress; it is again different when we have almost obtained the solution." (p.5) If an underlying structure is presumed to be inherent in a problem, then teaching problem solving can be viewed as mastery of a skill, as Polya did when he wrote that problem solving is "a practical art like swimming, or skiing or playing the piano: you learn it by imitation and practice." (Polya, 1962, p. v.)

Problems play an important role in the evolution of mathematical knowledge; their role in the teaching of mathematics is even more crucial. The common inclination to confer on them independent structure needs explanation. Even Lakatos after providing a rich and wonderfully social portrayal of the evolution of the Euler conjecture in *Proofs and Refutations* (1976) gives way to this inclination when he wrote:

> Mathematical activity is human activity. Certain aspects ... can be studied by psychology, others by history. Heuristic is not primarily interested in these aspects. But, mathematical activity produces mathematics. Mathematics, this product of human activity, 'alienates itself' from the human activity. It becomes a living, growing organism which *acquires a certain autonomy* from the activity which has produced it; it develops its own autonomous laws of growth, its own dialectic. The genuine creative mathematician is just a personification and incarnation of these laws which can only realize themselves in human action. Their incarnation, however, is rarely perfect. The activity of human mathematicians as it appears in history is only a fumbling realization of the wonderful dialectic of mathematical ideas. But any mathematician, if he has talent, spark, genius, communicates with, feels the sweep of, and obeys this dialectic of ideas.
>
> Now, heuristic is concerned with the autonomous dialectic of mathematics, and not with its history, though it can study its subject only through the study of history and through the rational reconstruction of history.

A footnote from Lakatos' students suggests that the Hegelian influence on him, apparent in this statement, was lessened by the influence of Popper, and thus this statement might have been significantly revised had Lakatos not died at such a young age. Nonetheless, his statement illustrates the strength of the human inclination to attribute to mathematical problems a timelessness and an independence from humanity.

IV. A CONSTRUCTIVIST VIEW ON MATHEMATICAL PROBLEMS

However, to confer on problems an independent ontological status as Lakatos seems to do in this passage and is so often done by problem solving advocates runs counter to the constructivist perspective. Problems, to problem solving advocates, are described as situations which require the solver to put together information in a new way. Polya describes problems as possessing an unknown (that which we seek), a given (that which we know as data which allow us to recognize the unknown) and a condition which will eventually link the unknown and the data. The problem in this description appears to be located in the very structure of the task.

The model that emerges appears to be as shown in Figure 1

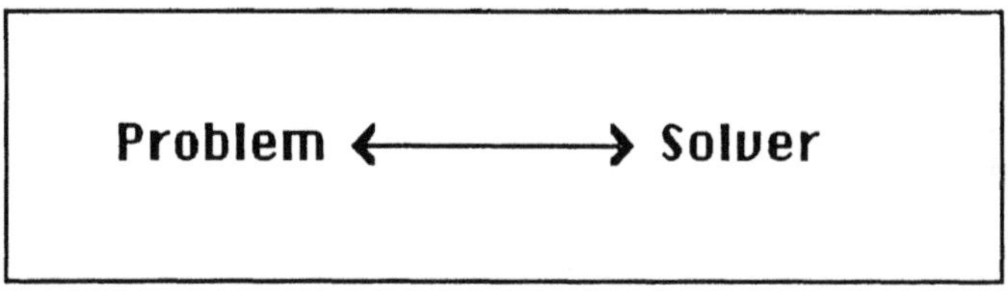

Figure 1.

Constructivists deny independent existence to problems. An example may demonstrate the point:

> Suppose a bank advertises 6% interest, compounded monthly, on savings accounts. If you deposit $100 on March 1 and withdraw it on April 1, how much interest will you have earned?

Without knowing that interest is, by convention, always quoted on an annual rate, one might easily answer $6.00. However, familiarity with this social convention allows one to deduce that 6% *per year* is equivalent to 0.5% per month, and that the correct answer is 50. The structure is not in the problem – it is in the socially and contextually defined meaning of the words as interpreted by the listener. As von Glasersfeld reminded us.

> As seasoned users of language, we all tend to develop an unwarranted faith in the efficacy of linguistic communication. We act as though it could be taken for granted that the words we utter will automatically call forth in the listener the particular concepts and relations we intend to "express" We tend to delude ourselves that speech "conveys" ideas or mental representations. But words be they spoken or written, do not convey anything. They can only call forth what is already there." (p. 485)

For the constructivist, the problem is only defined in relation to the solver. A problem is only a problem to the extent to which and in the manner in which it feels problematic to the solver. When defined this way, as a *roadblock to where a student wants to be*, a problem is not given independent status. In order to differentiate this approach to a problem from the typical use of problems in mathematics classrooms, I have chosen to use the term, *the problematic,* to refer to a student's "roadblock". In learning to listen to students, the constructivist devotes considerable time to imagining how the student views the problem.

In discussing the relationship between a problem solver and the problematic as s/he sees it, one still needs to explain why problems are so crucial in mathematics. If they do not "carry content", why are they so compelling? Most everyone will admit that it has been the examples, the problems, not the definitions and proofs through which they have learned mathematics – and this needs explanation.

The constructivist responds to this in two ways: by reconceptualizing what knowledge is and by describing the process of problem solving in relation to its affective character. In redefining knowledge, the constructivist draws on the fundamental processes described by Piaget: assimilation and accommodation as they act to restore equilibration. Human beings are living systems that seek equilibration and problems disrupt that equilibration. Once we admit of a problematic, (i.e. notice a disturbance), we work towards reestablishing equilibration. In this sense, problems while acting as roadblocks to where we wish to be are also "calls to action" – they get us poised to operate on a system.

We act through sensory-motor and cognitive operations. We use tools and previously familiar systems of representation. Then, we monitor the results of our actions to see if the problematic has been resolved and equilibration restored. This may end the sequence, lead to a reconsideration and perhaps alteration of the problematic, and subsequently a new cycle of action and reflection. When an action or operation is seen as repeatedly successful in our progress towards resolution, we set apart that action or operation by various processes of naming it, discussing it, objectifying it (Confrey, 1985) or creating it into a tool or representation for further action. This is the process of reflective abstraction (Piaget, 1971).

These intellectual processes are, for the constructivist, the source and the content of knowledge. Knowledge is not the accumulation of information; it is the construction of cognitive structures that are enabling, generative and proven successful in problem solving. Thus, problem solving becomes an essential intellectual act – not an enrichment, an occasional pastime. Figure 2 provides a simple model to describe this process of construction.

The objectification of problems, seen in these terms, reveals a human tendency which is itself a part of the problem solving process. We act as though problems (and their solutions) are out there. Michael Polanyi in *Personal Knowledge* (1958) emphasized that a person (a student or a creative mathematician) needs to believe a problem capable of solution and be "looking for it [the solution] as if it were there, pre-existent" in order to consider it a problem. Thus, inherent in our establishment of a problematic is a belief that it is capable of being solved. We may not possess the confidence or arrogance to presume that we will be the successful ones – what we are expressing is our belief that we will recognize the solution if we were to see it, though we may not be able to articulate anything about it. Our assurance is only in its *possibility* of solution – an assurance which easily leads us to talk as though problems were out there.

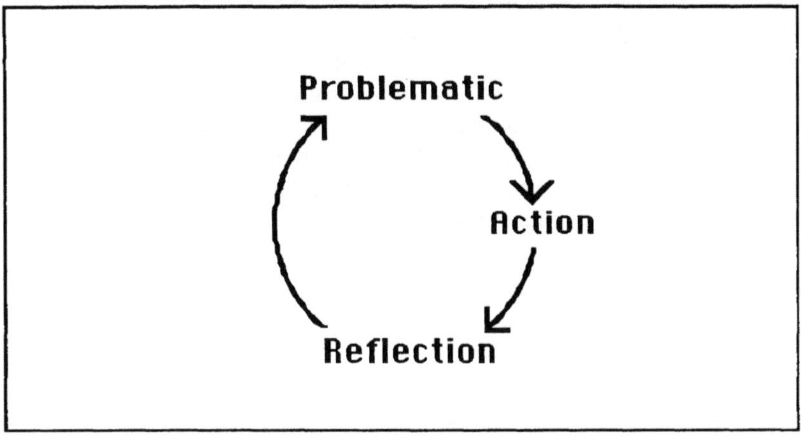

Figure 2.

His characterization of the emotional dimensions of problem solving explain further how our investment in the process encourages us to confer on the problem an independent existence. Our belief that "we can't create something from nothing" makes us wish to assume the problem solution came from outside. However, with a revised perspective on knowledge, we accept the recursive quality of knowledge construction and the emotional intensity helps to explain why problem solving is so successful in generating knowledge. As Polanyi put it, "obsession with one's problem is, in fact, the mainspring of all inventive power .... And the intensity of our preoccupation with a problem generates also our power for reorganizing our thoughts successfully ... (p. 127).

Polanyi summed up the power of the problem solving activity in his definition of a problem: "A problem is an intellectual desire ... and like every desire it postulates the existence of something that can satisfy it .... As all desire stimulates the imagination to dwell on the means of satisfying it, and is stirred up in its turn by the play of the imagination it has fostered, so also, by taking interest in a problem we start speculating about its possible solution and in doing so, become further engrossed in the problem."

This leads to a third assumpton for the constructivist position:

Asssumption Three: Problems serve a crucial role in the construction of knowledge. Problems reside in the mind of the student – not in textbooks or in the mathematics. Problems are felt discrepancies, roadblocks to where a student wishes to be and therefore catalysts for action. To accept a problematic an individual must believe that it is capable of being solved – and act as though the problem and solution were preexistent. The cycle of identifying (noticing) problematics, acting and operating on them and then reflectiing on the results of those actions is emotionally charged, motivating and demanding. It is this process of knowledge construction which is the critical site for constructivist researchers/teachers.

Solely defining a problem in relation to the solver does not prove adequate, for the constructivist researcher/teacher, who uses problems and tasks as the means of examining students' conceptions, promoting their continued development and assessing their progress. Thus, when problem solving is no longer a solitary affair, it involves an act of communication, an interaction which must be considered. The witness to the student's problem solving activity cannot claim access to some ontologically independent world of problems, so that s/he too has a problematic, an individual interpretation of the problem. Thus, the model must be revised as illustrated in Figure 3.

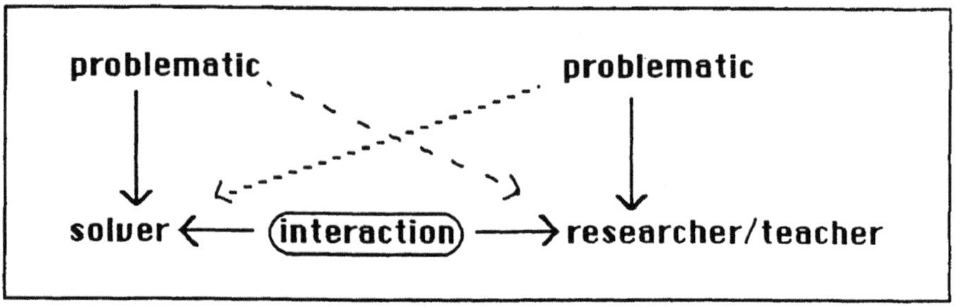

Figure 3.

Notice that lines drawn from each other's problematic are added to indicate that the researcher/teacher is trying to imagine what the student's problematic is like, while the student, in a classroom or interview, is trying to anticipate what the researcher/teacher wants. Over the course of the interview or the class, it is hoped that the two will come to believe that they understand better each other's expectations, conceptions and forms of solution – but this is best done when the student is engaged in his/her problematic and its resolution, and the researcher/teacher, guided by those responses, forms a powerful model and tests it while challenging the student with more sophisticated and yet relevant problems. Much of the success of the constructivist instructional or research model depends on how willingly the teacher 1) seeks to imagine how the student might be viewing the problem; 2) hears mathematical notions which differ from her/his own but possess internal consistency; 3) examines his/her own mathematical beliefs and 4) witnesses and describes the student's choice of operation (action) and method of evaluation and recording (reflection).

Assumption 4: Problem solving as enacted in interviews or constructivist instruction is an interactive process. The interviewer selects a task for its potential to invite students to engage with a particular mathematical idea. The task will yield to multiple interpretations and resulting approaches. The interviewer must seek out an understanding of the students' problematic, choices of actions and means of reflection. The interview setting will itself promote more self-reflection and a stronger approach to knowledge construction. The definition of the problem, of what concepts are related and of what constitutes an appropiate answer will evolve over the course of the interview.

V. STUDENT ERRORS AND MISCONCEPTIONS

A third interpretation of the first claim, that when students genuinely engage in solving mathematical problems, they proceed in personally reasonable and productive ways, would be to interpret it to refer to reasoned error patterns. Research on the diagnosis of student errors has indicated that overly quick and localized responses by teachers to errors can result in a failure to see how individual mistakes can be linked together and remedied more effectively as classes. Though this research has proven relatively effective in remedying errors on simple arithmetic computations (Van Lehn, 1983) its effectiveness in more complex circumstances has been limited.

A more promising interpretation of the phrase "personally reasonable and sensible ways" links it to the research on student misconceptions. Researchers have documented that students hold mini-theories about scientific and mathematical ideas that the theories and their forms of argument must be understood and directly addressed if students are to come to a more acceptable understanding of the concept. Unlike error patterns, these mini-theories relate formal scientific/mathematical meanings for terms with their everyday usage, examine how the theories relate to historical development and discuss how the theories reflect the child/student's view of science or mathematics as a whole. (For reviews of this literature, see Confrey, 1990; Perkins and Simmons, 1987; Driver and Easley, 1978)

Since many of the researchers in this area have contributed significantly to the constructivist program (Pope, 1985; Driver and Erickson, 1983; Novak and Gowin, 1984) no general distinctions can be offered. However, debate within this community over what to label the student's perspective is an indicator of the variation which exists in people's interpretation of constructivism. Labeling a student's model as a *misconception* fails to take in consideration the perspective of the student, for whom the belief may explain all instances under consideration and fail only in cases to which s/he is not privy. Another optional label has been that of alternative conceptions which stresses the possibility of multiple solutions, but subtly seems to give them a lower status, "alternative" as opposed to "normal" conceptions. Finally, others have chosen more simply conception, which omits any indication that the perspective may deviate considerably from the expert's position. For the constructivist, this difficulty can be resolved by using the term "conception" while always declaring a frame of reference (observer, expert, or participant) and indicating whether it seems adequate from that person's perspective.

VI. A CONSTRUCTIVIST VIEW OF ERRORS

In my interpretation for constructivism, the phrase "personally reasonable and sensible ways" needs to incorporate certain commitments. As we examine students' conceptions and encounter responses which deviate from the traditional presentation, we need to keep in mind that:

1) seldom are students' responses careless or capricious (Ginsburg, 1977). We must seek out their systematic qualities which are typically grounded in the conceptions of the student. If we **assume** this, we will increasingly understand the sensibleness of their approaches from their point of view. Thus, our first efforts must be towards encouraging students to describe their beliefs so that we might explore them.
2) frequently when students' responses deviate from our expectations, they possess the seeds of alternative approaches which can be compelling, historically supported and legitimate if we are willing to challenge our own assumptions.
3) these deviations provide critical moments for researchers to glimpse and then to imagine how students are viewing an idea. The role of the discrepant event in building up conceptual structures is as essential as refutation is in the conduct of science/mathematics. It is essential in an individual's progress in the development of the ideas and also provides an opportunity for us to describe the student's conceptions. When the student proceeds in accord with our own models and expectations – we assume agreement with our models, but as von Glasersfeld (1984) reminds us, conceptual frameworks must *fit* our experience, as any key with the appropriate appendages can trip open a lock. That key need not *match* our own. It is at points of contact, at moments of discrepancy, that we have the highest probability of gaining insight into another person's perspective.

Assumption Five: Students' responses which deviate from our expectations as researchers/teachers can appear to be reasoned and well-considered to the student. They may be entirely legitimate – as an alternative perspective, or be effective for a limited scope of application. We must encourage students to express their beliefs, keeping in mind that deviations provide precious opportunities for us to glimpse the students' perspectives.

VII. AN EXAMPLE: SUZANNE

In order to demonstrate how one learns to listen to a student in a constructivist tradition, I have written up an episode which occurred over two interviews with a college student. The questions listed below summarize the major issues raised in the previous sections and provide a framework for the reader to consider as s/he considers the example:

1. What mathematical ideas are under investigation and how can they be viewed as genetically developmental, functional for making sense of experience and viable in solving problems?
2. What problematic is the student undertaking and does she solve it adequately? From whose perspective?
3. How does she act to solve her problematic? What operations, representations and tools does she use?
4. How does she reflect on her actions in relation to the problem?

TIME-LINE PROBLEM:

The world is very old, and human beings are very young. Significant events in our personal lives are measured in years or less; our lifetimes in decades; our family genealogies in centuries; and all of recorded history in millennia. But we have been preceded by an awesome vista of time, extending for prodigious periods into the past, about which we know little - both because there are no written records and because we have real difficulty in grasping the immensity of the intervals involved.

- Carl Sagan, Dragons of Eden

The problem: Represent the following dates on a number line:

Period or epoch	Years ago	Development of life on earth
now		Development of science and technology
Renaissance	500	Voyages of discovery from Europe and Ming Dynasty, China
	1000	Mayan civilization; Crusades; Sung Dynasty, China
	1800	Zero and decimals invented in Indian arithmetic; Rome falls
	2000	Euclidean geometry; Archimedian physics; Roman Empire; birth of Christ
Holocene epoch	10,000	development of agriculture
	500,000	Domestication of fire by Peking Man
Pleistocene epoch	2×10^6	Modern human beings develop; mammoths and woolly rhinos flourish
Miocene epoch	2.4×10^7	apes, bats, monkeys, whales
Oligocene epoch	3.7×10^7	rodents, cats, dogs, elephants, early horses
Eocene epoch	5.8×10^7	birds, amphibians, small reptiles, fish
Paleocene epoch	6.6×10^7	flowering plants; small mammals
Cretaceous period	1.44×10^8	dinosaurs with horns and armor common
Jurassic period	2.08×10^8	dinosaurs reach their largest size
Triassic period	2.45×10^8	cone-bearing trees plentiful; insects; appearance of turtles, crocodiles, dinosaurs
Devonian period	4.08×10^8	the first forests; many fish and amphibians appear
Ordovician period	5.05×10^8	Trilobites, corals, and shelled animals
Cambrian period	5.7×10^8	fossils plentiful for the first time
Precambrian time	1.1×10^9	coral, jellyfish, and worms
	3.5×10^9	Bacteria
	4.5×10^9	no living things are known
Big Bang	1.5×10^{10}	

Figure 4.

The preceding example is drawn from a teaching interview which I conducted on students' understanding of exponential functions. The student interviewed was a college freshman at Cornell University who was majoring in nutrition. At the time of the interviews, she was taking a remedial algebra/trigonometry course. She volunteered to participate in a study of students' mathematical reasoning and was paid for her participation. She met with the interviewer twice weekly for one-hour sessions over a five week period. The task described took place during the second and third interviews. The task (reproduced in Figure 4) was to represent a series of dates of significant historical events given in scientific notation (for the most part) on a number line. Scientific notation refers to the system of writing numbers as a product of a number between 1 and 10 and a power of ten. The task was designed in order to examine how students understand scientific notation and, in particular, the relations among the exponents as orders of magnitude.

Suzanne was given the problem statement and asked to read the introduction out loud and "to represent the dates on a number line". She was provided a calculator, ruler, pens, pencils, a stack of computer paper and pads of graph paper. She chose to use the graph paper to make her number line and began by stating that she will "divide it up" and go from $1.5 \cdot 10^{10}$, $1.5 \cdot 10^{9}$. When asked what she is looking at to make that decision, she says "Well, um, let's say I put in 1.5 times 10^9 here, if I was going by thousands of years or whatever it is. Um, I would have to fit in here 3.5 whatever it is.." She proceeds to mark off her intervals, $1.5 \cdot 10^{8}$, $1.5 \cdot 10^{7}$ etc.

This decision immediately meant that these intervals were spaced as in a logarithmic scale: as she moved right to left, each equal-sized interval spanned 10 times as many years as the previous interval. For example the interval from $1.5 \cdot 10^{1}$ to $1.5 \cdot 10^{2}$ covered from 15 to 150 years ago whereas $1.5 \cdot 10^{2}$ to $1.5 \cdot 10^{3}$ covered from 150 to 1500 years ago or 10 times as long (Figure 5).

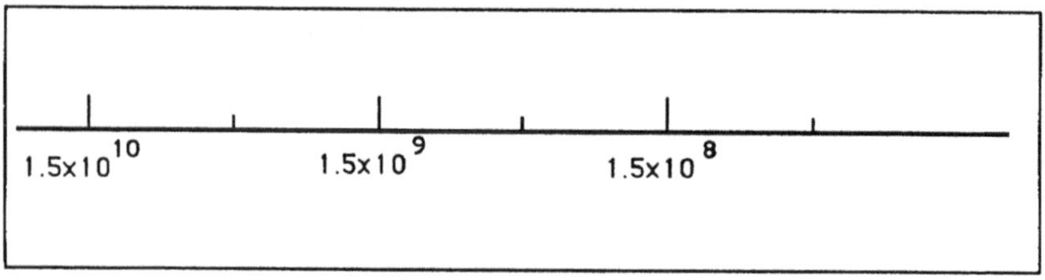

Figure 5.

In order to decide on the spacing in terms of boxes between the intervals, she counted the number of boxes across the graph paper and the number of events which are listed in the chart (19 events) and divided number of events into the number of spaces (47 spaces). She indicated that "..I'm probably going to need a calculator because you mark this at the proper divisions between each one."

From this, she decided to space her labels two squares apart.

In one respect, it seems that Suzanne is muddled. She has not proceeded in the textbook fashion of discerning the range and then choosing a scale, labelling it, and then placing her points. Her method of deciding on the spacing seems flawed and her decision to make the intervals in increments labelled $1.5 \cdot 10^n$ seems to demonstrate a failure to understand the impact of exponents on the size of numbers. It seems unlikely that she will complete the task successsfully.

However, if we struggle to listen to her as a competent problem solver, she has also provided us a rich opportunity to glimpse what constructs she is using in creating such a number line. Her first concern was with fitting the data on her paper. She was aware that "I can use the whole room if I want to, because this is such a long period of time," and she recalled an exercise in ninth grade earth science where they created a timeline with pictures. However, she decided to limit her task to one piece of graph paper, and worked towards mapping the data within those restrictions.

The interviewer who is charged with hypothesizing a model of her problematic listens carefully to her words to guess at her task – she saw her task as "dividing up" the page (creating intervals) and "fitting in the points" (locating all the points on that page). She stated her own problematic as "getting them all on, from first to last." It seems that her immediate task was to place the points on the scale. The concern for representing the relative passage of time was never mentioned early in her activities.

Within the constructivist paradigm, one would seek to imagine how these concepts, range, interval labelling and spacing within the interval are understood within the context of her problematic. We see that our "conventional wisdom" on how to construct a number line does not apply. She does not proceed in textbook fashion by: 1) identifying units (the variable in the one-dimensional scale), 2) determining the range, 3) choosing an origin and 4) deciding on an appropriate scale. Her problematic is to get the points on the line, i.e., to find their locations.

To do so, she selected the Big Bang, the first point in the set, and hence ended up with a marker of $1.5 \cdot 10^n$ a choice which she will revise later. She chose her markers to proceed consecutively, but the counters she relied on were exponents; she exhibited no evidence that she recognized this implied a change in interval size as she moved right to left. It felt orderly and it allowed to cover her range of points systematically. From her later actions, it seems reasonable to assert that in the early stages, she was unaware of the discrepancy between her scale and a linear one.

An important insight for the researcher is that it seems that her construction of the number line (on which she later must place her points) is not clearly differentiated from locating the set of numbers (data) she must place. We witness this in her decision to use the Big Bang as her origin, despite its somewhat awkward value ($1.5 \cdot 10^{10}$), rather than to choose a value like $1 \cdot 10^{10}$. Another example is evident in her assumption that the events themselves will be equally spaced, rather than the intervals on the scale. Recall how she decided how many spaces to put between each label of a marker; she counted the number of events, divided 44 by 19, and spaced by twos to "fit them all on."

For the researcher, these deviations from our expectations raise interesting questions about the conception of the number line, since she must create this representation in order to place the points. We explore our own understanding of the interrelationships among scale, range, unit and interval (Goldenberg, 1988). Her initial lack of distinction can draw our attention to ways in which our own mathematical language makes such a distinction difficult to articulate. For instance, we refer to the representation of the number line without points labelled as a number line, yet, possess no language for referring specifically to a number line with the data points marked. In contrast, in Cartesian graphing (a representation of two perpendicular number lines), we ambiguously use the term, "graph" to refer to both the particular curve drawn (the relation) on the coordinate axis (and can put multiple graphs on the same axes) and to the combination of the particular curve and the coordinate axis. We do not refer to a naked coordinate axis as a graph.

As we consider these as two related tasks, we begin to view the problem as the coordination of two kinds of representation – numbers represented in scientific notation and numbers represented as locations of points on a number line. By not differentiating these systems of representation initially, she uses particular points and intervals from her list of events to build her number line and unknowingly breaks some of her tacit assumptions about numbers. Interestingly enough, as the task proceeds, and she most locate a variety of points in relation to her interval labels, she progressively solves local problems and ends up with a practical solution.

Her second attempt has her using two sheets of paper, and she increased her interval size to four boxes. She did so in response to beginning to try to plot the three events in Precambrian time and suspecting that she lacked the necessary spacing. As she began to imagine plotting points, she decided she has made a mistake. She moaned, "Oh, I totally messed this up, because when I counted I didn't take into account that there are four, five, six at 10^8 and so many at 10^7 so that's not going to work .... I'm going to end up with a bunch of extra marks with nothing to put there." She continued to examine the chart and then struggled to create a language to describe her error. She said, "So when I counted the total [number of events] I included them [multiple events with the same exponent] in the total, not thinking that they are not necessarily in the total but they're in the **group** .... So, what I have to do is, say, count the exponent as the group and not each individually."

The significance of this event for the constructivist is twofold: 1) she was easily convinced to change her mind on this, because she encountered conflict and acted to resolve it; and 2) her development of a language by which to convey to the interviewer her insight required significant effort and probably helped to stabilize that insight. In this regard, the interview setting constitutes a different setting for learning than would likely occur if she had been alone. The use of the term, group, is an interesting one, for it establishes a way of reducing the size of her problematic of mapping from scientific notation to the number line – each point is a member of a **group** as defined by its exponent, and that exponent corresponds to a single interval on the number line. She now must only determine how to map within the intervals. Thus, her problematic has changed as a result of her own actions.

In the next exchange, Suzanne acted on her insight to determine how many intervals she needed. Since the final seven events are not presented on the table in scientific notation, she transformed them. She proposed to change 500,000 into $5 \cdot 10^5$ counting the zeros but expressed a lack of security with scientific notation. When the interviewer responded by asking, "What is 10^5?", she multiplied it out to get 100,000 and checked it.

Her encounter with the label "now" provided her with another opportunity to differentiate between the representation of the number line and the placement of her particular data. It also drew her attention to her units. She stated, "And I guess I am going to just put that on the graph [the word now]. I can't really. I can put it to the zero. I mean, I guess, Oh, I can just, if I put 19[87]. Wait, this is actually zero, isn't it, because it's years ago? It's right now, it's because its zero years ago, so will go, um, I'll have to put something in here, as 10^1 just so the graphs will stay even ... it will be 1.5 times 10^1, so that, no, that may not be important as far as the data, but I need it to keep the intervals in my graph." This is a significant step for her for she makes the claim that the number line must meet a certain criterion, completeness, that may not be directly required by her data.

Notice that the placement of "now" cannot be represented using her geometric progression of exponents since $1.5 \cdot 10^0 = 1.5 \cdot 1 = 1.5$ and were she to continue her pattern of intervals, $1.5 \cdot 10^{-1}$, $1.5 \cdot 10^{-2}$, $1.5 \cdot 10^{-3}$ which yield .15, .015, .0015. No extension of such a sequence reaches the value of zero. Because she is still primarily concerned with placing points, she sees her task as finding a place to put "now". She puts it on the point one, calling that "now".

This example illustrates how a student can simply dismiss an anomaly (and fail to "see" it), leaving intact the operating conception. The interviewer asked her if zero can be written in the same form as the other numbers, challenging her to see that it provides a challenge to her framework. She responded, "Um, I don't know. As $1 \cdot 10^0$ or 10^0 is 1, so it would just be one. I guess I could call it [now] one if I want to ... since it's my graph." On hindsight, it is unfortunate that the interviewer did not pursue this either immediately or later.

Suzanne now set about creating the third version of her number line. She counted to see she needed 11 intervals and decided to try a spacing of 10 boxes per interval. When that came out to be too large for her graph paper, she decided to use 8 squares. She drew this fourth version and labelled the intervals, $1.5 \cdot 10^{10}$, $1.5 \cdot 10^9$, $1.5 \cdot 10^8$ ... and wrote in zero for her last entry. She then changed her mind again as she faced the task of locating points and said "Actually, I shouldn't have used eight, because its hard to break that up. Probably 5 or 10. If I stayed at 10, it would have been a lot easier." Her choice of using ten may have signaled either an awareness that she was facing the task of placing decimals or that she was accustomed to seeing ten spaces in previous work.

She decided to continue to work with eight squares and encountered her next obstacle. The final resolution of this obstacle provided her with a consistent, albeit alternative, representation for this problem. She tried to decide how to divide the interval up. Her first suggestion was to divide the interval next to $1.5 \cdot 10^{10}$ into three parts and label tentatively from left to right the first section $1.0 \cdot 10^9$ and the next $.5x\ 10^9$ (Figure 6).

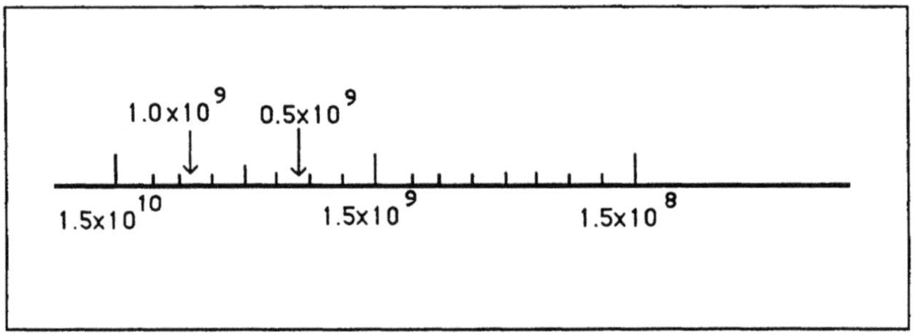

Figure 6.

She quickly stated her confusion. After a short pause, she concluded, "I am going to make this one [the marker] $1 \cdot 10^{11}$ and this, $1 \cdot 10^{10}$ so I can use "1" and have some place to put this [$1.5 \cdot 10^{10}$]. So, I can put $1.5 \cdot 10^{10}$ here." She pointed to the middle of the interval (Figure 7).

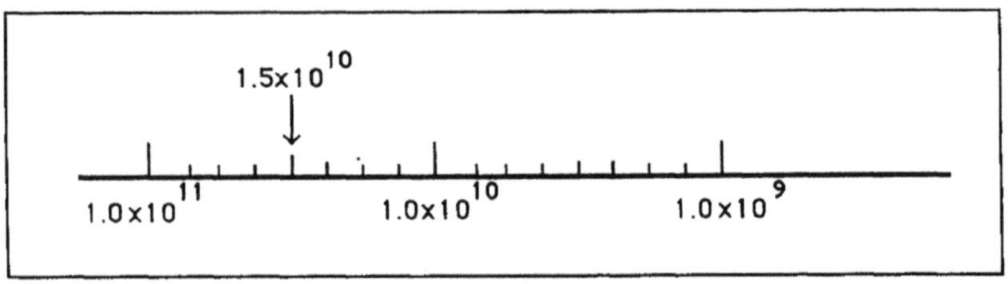

Figure 7.

This change in labelling the intervals can be viewed as another step in differentiating the formal representation of the number line from the particular data. It also provides an opportunity to glimpse some of her assumptions which are in operation. She assumes that .5 must represent a "middle" point, perhaps because it is so frequently the middle in number lines of consecutive integers; perhaps because she is attending to the consecutive values of the exponents. If she had considered her intervals as 10^n and 10^{n+1} (rather than as $1 \cdot 10^n$ and $1 \cdot 10^{n+1}$), she might have claimed that the middle is $10^{n+.5}$, a claim which might have led towards a logarithmic scale. However, given her revised problematic, locating points within the intervals, she is most likely to be attending exclusively to the decimal number (n.nn) rather than the exponential (10^n).

Because she remained dissatisfied with having eight rather than ten squares, she redrew her number line with the interviewer's assistance. She then had the number line shown in Figure 8. It started with $1 \cdot 10^{11}$ and proceeded downward by powers of ten till $1 \cdot 10^1$ next to which was placed a final interval with a marker labelled NOW. She chastised herself stating, "My biggest problem is I don't think things through enough before I start."

Her first inclination was to split each interval in half, which later she explained was to label it 1.5, but she caught herself and said, "no, I'm not going to do that. I am going to ... since our 10 ...". She counted off the spaces between her markers, starting from her right marker at 1 and moving to the left counting 2, 3, 4, 5, 6, 7, 8, 9, 10 .... This caused her to reach 10 on the ninth square and to recalculate that value and say, "no, if I do 10 times 10 is ..."

Suzanne was experiencing the impact of a variety of competing frameworks. Her markers read $1 \cdot 10^8$ and $1 \cdot 10^7$. They seemed well-ordered to her, but she needed to figure out how to divide up the interval. Her first framework was to hypothesize that $1.5 \cdot 10^7$ would be halfway in between – a theory based on the consecutive values of the exponents. Her other framework guided her to rely on her choice of ten spaces and its resemblance to the decimal system and it led her to count off the values. This method, however, produced the surprise result that she ended up with an extra space since her count begins at one.

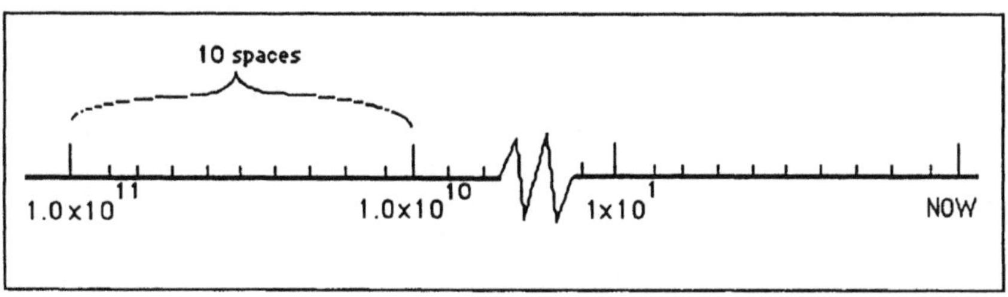

Figure 8.

She admits "Um, I am confused now as to what I'm doing. Um.." and then chose to write out the numbers in standard form below the scientific notation. On the interviewer's request, she restated her problem, this time using the standard form of the numbers. "I don't know how to go from 100 million to 10 million. It's just 10, 10 million times 10 is 100,000. The scientific notation is messing me up. This would be 10 million and 500 thousand, right? [She pointed halfway in-between 100 million and 10 million.] It would be halfway in-between, right? So it would be 1.5 right here?"

The interviewer repeated her claim asking if 10,500,000 is halfway between 10 million and 100 million and she quickly responded that it's halfway between 10 million and 11 million. Although she was confused, by writing the numbers in standard form, she has provided herself a third representation (in contrast with her segmented number line and her scientific notation) and she used this one to resolve her dilemma (Figure 9). She said, "All right, then I have too many spaces ... since this is 10 million and this is 100 million, I have to go by 10 millions. So, it would be – if this is 10 million – 20 million, 30 million, 40 million, 50 million, 60 million, 70 million, 80 million, 90 million. Then I'd have an extra box." Then she switched back to scientific notation and said, "then it would be $2 \cdot 10^7$, $3 \cdot 10^7$, $4 \cdot 10^7$ – like that. On interviewer's request, she confirmed her method on the interval from 1 million to 10 million counting off by millions. As the interview ended, she expresses her need to redraw the number line and the interviewer agrees to provide one with nine divisions per interval for the next session.

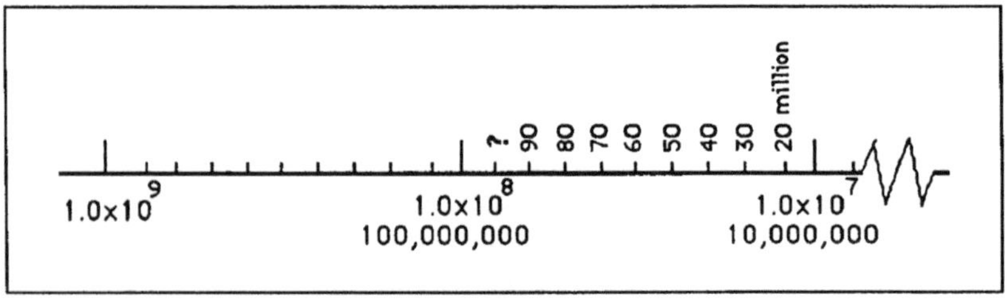

Figure 9.

To this point, Suzanne has created a number line which resolves all her outstanding dilemmas. She has created a number line which involves two mapping procedures – the first maps the events within a group by their decimal numbers and to place them on a scale ranging from 1.0 to 9.9 ... within that interval. In more formal terms, her proposal creates a complete mapping from the set of all real numbers greater than zero to the number line representation. It is a combination of a logarithmic and a linear scale.

The interviewer began the next session by asking her to review what she had done to which she replied: "I tried to make a number line with the numbers given here [in the chart], and I put them in scientific notation, just to make it easier ... I tried to make even divisions so I could divide them up, I did them by tenths. I started 10^{11} and 10^{10}, 10^9, and I put in finally, once I figured them out, put the proper numbers and space in between so that each could be labelled evenly, like $1 \cdot 10^{11}$, $1 \cdot 10^{10}$, down like that. From there it would be easier to divide them up like into the decimal point and things like that."

Curious to see if the solution she has proposed will satisfy the overall expectations Suzanne has for the task (recognizing that it does solve her local problematics), the interviewer asked Suzanne what would constitute a successful completion of her task. She identified three criteria: 1) to be able to locate the points on the chart; 2) to be able to look at the number line and "clearly see what was being, was trying to be said, or trying to be explained by things in the number line"; and 3) to see the years passing, "the difference between them. When you look at the line, you can see clearly that this is obviously closer. This one point is closer to this point than this one is. So there is a longer period of time between here and here."

Her answer is striking, because her number line only satisfied her first criterion completely. To this point, we see no reference by her to the context of the problem, the passage of time; and yet, when asked she states this as a criterion. Later in the interview, she learns to "see" on her number line and to calculate the years passing - but the placement of numbers remained of primary concern.

She then located the events listed on the number line with little difficulty. Her representation allowed her to map the dates for the events listed in an algorithmic fashion. The interviewer began to raise questions to cause her to consider the impact of a changing unit size across the intervals. The question posed is "did the dinosaurs live for a longer period than humans, or have humans lived longer?" They decided to assume that the dinosaurs died out during the Triassic period. Thus it turns out they lasted approximately 100,000,000 years and people, having entered the timeline at 2,000,000 years ago, have survived a much shorter length of time. On her number line, by distance considerations alone, it appeared that humans have outlived the dinosaurs (Figure 10).

She answered, "They began a longer time ago, but they lasted a shorter period of time". This is the response one would expect her to have given, based on the appearance of her number line.

The interviewer asked to be convinced, and she responded at first by saying, "Each exponent is a million years. I guess that's what we were doing. Is um, what did we say. I have to think about this for a second." She wrote out the standard notation for all of her intervals. Then she said, "OK, each of these, each of the divisions signifies a multiplication by 10 um? each of the large division. This is 10 years multiplied by 10. 10 million multiplied by ten is 100 million. With considerable effort, she pondered the length of the interval for the dinosaurs which we had marked as from $2.45 \cdot 10^8$ to $1.44 \cdot 10^8$. She quickly said it's one and with a long pause added, "I guess it would be one … hundred million." To work this problem, she stayed within her number line representation, switching from scientific notation to expanded decimal form. In then deciding how long humans had survived, she also used her number line and began to add intervals. She added the first interval correctly (from $2 \cdot 10^6$ to her marker, $1 \cdot 10^6$) as 1 million but then called the next interval 100,000 (rather than 900,000) and then simply concluded that dinosaurs were around longer.

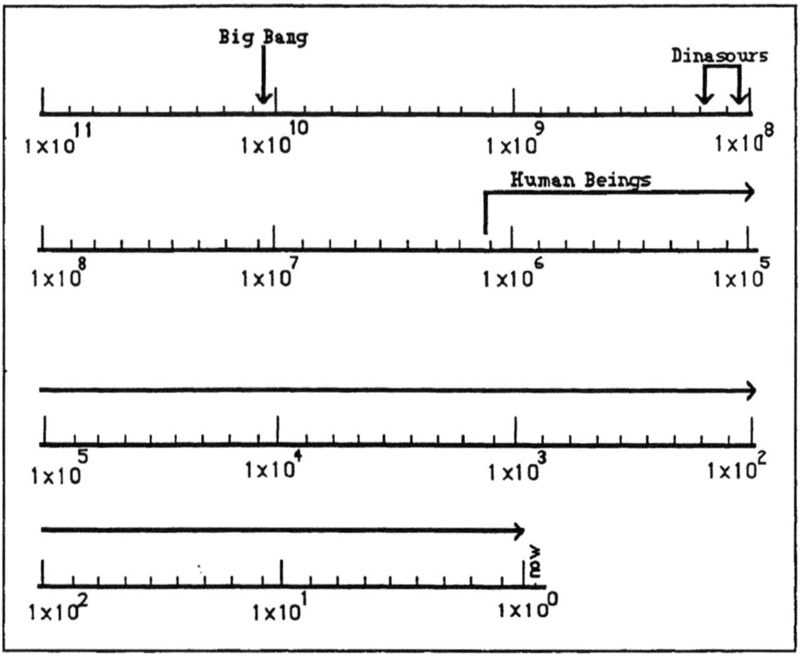

Figure 10.

By asking this question, the interviewer hopes to stimulate Suzanne's thinking about interval size. Whereas before we have heard Suzanne express her awareness that the value of numbers between intervals increase ten times, "each of these, each of the divisions signifies a multiplication by 10", we saw no evidence that she was interpreting this in relation to the passage of time as represented by the length of the line. By working across her forms of representation, the length of the line, subtraction of scientific notation and subtraction of expanded format, she gains some insight into these connections. It is at this point that she begin to apply her last two criteria to her number line, to "clearly see what was being, was trying to be said" and to see the years passing. Once again, when presented with a problem, she acts to solve it in an appropriate way.

Her decision to determine how long humans had been around by adding her intervals, 1 million + 900,000 + 90,000 + 9,000 + ... (which would be an infinite process in her representation if she had arbitrarily located one as now) rather than by reading off the number 2 million provides an interesting insight into her understanding of the meaning of the labels on her intervals. It might have also led to an interesting inquiry into her awareness of the differences in her number line from the traditional linear one. However, it was not pursued.

Suzanne seemed to be reasonably capable of using her graph and seemed to understand what changes occurred in the value of the boxes. She summarized by saying, "Each box is worth a lot more over here than it is over here; you multiply by ten." To pose a final test, the interviewer asked her how long it took for a human being to discover fire, a calculation which required her to calculate between two intervals. She responded, "Oh, um, 500,000 right here [counting the units in the smaller interval]. So it's 500,000 thousand right here. And then right here is a million so it will be 1,500,000." She used her representation quite successfully by breaking the problem into two pieces. She calculates each interval piece separately using its correct unit and adds the total.

It appears that Suzanne has created a logically consistent solution to the problem. Once she learns to use it, it satisfies all her criteria and allows her to complete the task. The question for discussion is how legitimate is her solution?

VIII. DISCUSSION

The episode described is not unusual. Nearly every extended teaching experiment yields inventive solutions such as the one that Suzanne has demonstrated. Regardless of the extent to which one finds her solution mathematically acceptable, it seems reasonable to point out that she has certainly engaged vigorously with the problem and demonstrated the ability to create sensible solutions to each obstacle she encountered.

It seems also reasonable to conclude that Suzanne seemed relatively effective in her use of her representation. She recognized the changes in the intervals and could adjust to account for those changes. Furthermore, it seems fair to claim that she does solve most of her problematic. From her perspective, what she has created is a perfectly adequate solution.

This is not to suggest that no further work might be done with Suzanne on these types of problems. She does not appear to have engaged deeply with the implications of the absence of a point for zero – an implication which might be important for contextualized problems. The tension between seeing the label as a magnitude from zero (as in the two million years of humanity) and as a sum of an infinite geometric sequence is unresolved. And, in future work with all negative exponents in a comparison of small-sized objects, we witness an entirely different approach. Finally, she does express her own uneasiness with having different sized units and her wish that she could have found a solution which did not demand that "concession".

Secondly, her solution path provides us with interesting and useful information about the psychogenesis of a number line. It suggests that in coordinating two representations, scientific notation and a number line, we witness a process of progressive differentiation. She does not treat the scaling of the number line as her prior task and the placement of the points as its successor. The two acts are intertwined and slowly differentiated as she proceeds. The advantage of this process despite its inefficiency from the expert's perspective is that she begins to create a functional need for ideas like origin, range, units and scale. With one experience, it is doubtful she could articulate these constructs in any sophisticated way, but it might turn out that she could use these ideas successfully in future tasks similar to this one.

The question which remains is whether to consider the work she has done is legitimate mathematics. Many would deny its acceptability, because it is a hybrid between logarithmic and linear scales. Some would argue that it does not show the passage of time comprehensibly. However, if this objection were taken as valid, one would have to reject the logarithmic scale as well. Furthermore, if she were to create a fully linear scale with one inch equal to 10,000 years, such a representation would be fully fifteen miles long.

Others would deny her solution legitimacy by claiming that her solution is a distortion of a logarithmic scale and therefore wrong. If one were to develop a purely logarithmic scale given her method, one would change all the numbers into a form of $10^{n.nn}$ with no decimal number up front and then place them on a scale based on the value of the exponents. Thus, $10^{.5}$ would be halfway between 10^0 and 10^1 and thus the midpoint between 1 and 10 could be $\sqrt{10}$ or approximately 3.16 rather than 5.5 on her nine point scale (Figure 11).

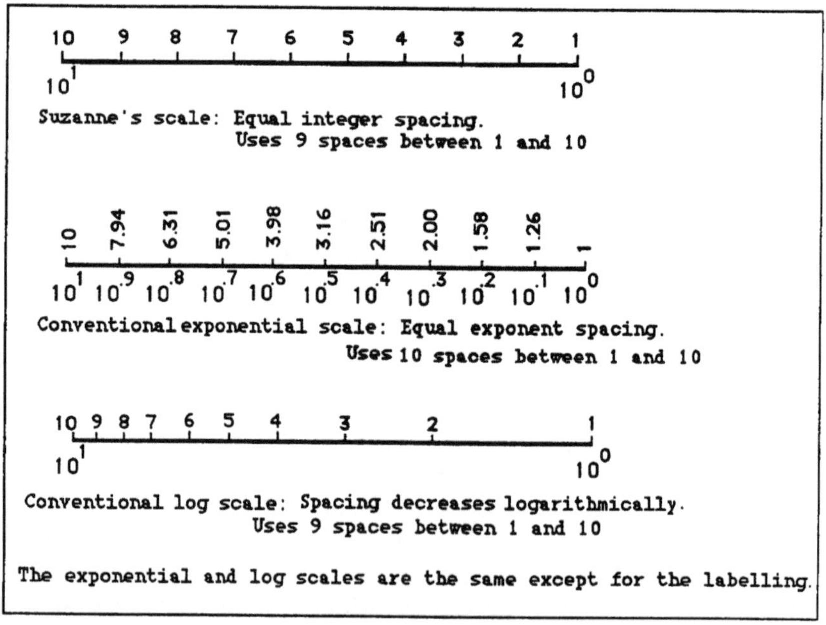

Figure 11.

Her scale is a distortion of the logarithmic scale, but its rejection as legitimate mathematics does not follow. Just as one would not agree to reject the logarithmic scale because it is a distortion of a linear scale. Perhaps we would be better off to refer to the changes in scale as **transformations** and examine their advantages and disadvantages. For the constructivist, however, no rejection of the legitimacy of her method can be made simply on the basis that it does not mirror the conventionally accepted solution. This may be a disadvantage in communicating her solution to others, but it does not undermine the valdity of her solution.

I would claim that Suzanne's solution is: 1) mathematically valid within a specifiable system of mathematics, 2) preferable to a logarithmic scale for displaying data in scientific notation, and 3) reveals a fundamental tension between multiplicative structures and counting structures which is built into our systems of numeration in fundamental ways, and hence her solution provides us as researchers with an opportunity to reconsider our own understanding of numbers.

1. *Mathematical validity.* Suzanne has created a system with a one-to-one mapping between the set of all real numbers greater than one and the number line. On her mapping, she identifies one as her origin (and decides to call that now – a relatively small quirk when we are spanning 10^{10} years.) She develops a set of algorithms for working, identifying the units in any given interal as 10^n and for adding and subtracting within an interval. Her algorithm for crossing intervals is to split it into two parts and treat each one separately. Such a system lacks a constant rate of change either additively (as in linear scales) or multiplicatively (as in logarithmic scales). It does possess a predictable rate of change, which is a combination of the two.

2. *Better for Scientific Notation.* If one wishes to claim that Suzanne's scale is distorted (in a negative sense), one would have to ask distorted from what? For the constructivist, there is no "real" number system, preferable to all others. Number systems are created to allow us to accomplish certain things, to compare, represent, measure, order etc. Fractional form is easier when dividing up quantities; decimals allow easy comparisons of size. Thus, if her number system is a distortion of a logarithmic scale, then scientific notation would be also a distorted system. Few would make such a claim.

In fact, I would claim that her system of representing those numbers on a number line creates the best imaginable fit with scientific notation. As with scientific notation, she is able to represent a huge span of time in an efficient manner. Her two part mapping allows her first to identify the interval and then to locate the decimal number. Scientific notation is designed to do exactly the same thing. By forcing the decimal portion into a standard form between one and ten, the scientist can compare the orders of magnitude. S/He can quickly reduce the problem to one of working with integral values, our counting numbers and our most familiar system.

There is a functional elegance to scientific notation. It is a transformation of a logarithmic scale, but a useful one. Suzanne has simply created a number line with the same advantages and disadvantages. In order to add and subtract, you must have like exponents in scientific notation. In Suzanne's system, you must break the problem into like units. By creating her system, Suzanne has given us an opportunity to consider thoughtfully the assumptions behind scientific notation and behind the construction of the number line.

If mathematics is viewed a functional, the emphasis is not with mirroring some unknowable reality, but in solving problems in ways that are increasingly useful in one's experience. In mathematics that means one must weave together multiple representations in order to solve interesting problems. Effective uses of multiple representations possess two characteristics: 1) they display enough variance across representation to allow insight, and 2) they allow enough convergence among the representations to give one confidence in one's solution. Suzanne's coordination of the representations of number lines, tabular data, scientific notation and expanded form illustrate these two characteristics quite nicely.

IX. MULTIPLICATIVE AND COUNTING STRUCTURES

Suzanne's treatment of this problem provides insight into a fundamental tension in mathematics: the relationship between multiplicative and counting (additive) structures. Confrey (1988, 1990) has argued for an existence of two relatively independent primitive structures in young children: counting and splitting. She has suggested that when these primitive structures are assimilated to each other (when we define multiplication prematurely as repeated addition) and can cause certain conceptual obstacles (powerful ideas which are difficult to learn) emerge. Place value is an example of a conceptual site in which students must coordinate their use of *values* ranging from 0, 2, ..., 9, and *placement* into powers of ten which are multiplicatively increasing.

Suzanne's solution and the insight which it gives into the structure of scientific notation provides another illustration of the uneasy but useful coordination of multiplicative structures (her intervals) and counting structures (within intervals).

X. CONCLUSIONS

The paper began with the statement of two assumptions concerning the application of constructivist theory to mathematics education. It made the claim that these two assumptions needed to be interpreted in a particular way in order to be compatible with constructivist epistemology. These included the claims that the mathematical content must be viewed as a human creation designed to serve human purposes; that the mathematics created inspired confidence because of its interwoven representations and that in any circumstance one should seek multiple mathematical solutions, diversity rather than uniformity.

In the second section, I discussed the idea that within the constructivist paradigm, problems must undergo individual interpretation and labelled this the problematic. I sought to argue that to understand a student's actions, one must seek to model their problematic and not presume it is identical to one's own. I further argued that apparent errors, discrepancies from the observer's expectations, provide particularly useful opportunities for one to imagine what a student's problematic might be like. Finally, I argued that in examining students' problems and methods of solutions, one has an opportunity to reconsider the mathematics involved.

Suzanne's solution to the number line problem provided such an example. She created and solved a series of problematics each of which was reasonable though somewhat atypical. Her solution was unconventional but inventive and original. She demonstrated adequate insight into it. For the most part, deviations which appeared erroneous from our position of expertise turned out to be surmountable conceptual obstacles for her. When given an opportunity to engage in her own cycle of problem solving, she demonstrated considerable progress, insight and experienced, by her own report, a sense of satisfaction.

NOTE

1. This research was funded under grants from NSF (N. MDR-8652 160) and the Apple Corporation.

BIBLIOGRAPHY

Bruner, J.S. (1960). *The Process of Education.* NY: Vintage Books.

— (1966). *Toward A Theory Of Instruction.* NY: W.W. Norton.

Confrey, J. (1985). "Towards A Framework For Constructivist Construction". *Proceedings of the Ninth International Conference for the Psychology of Mathematics Education.* Noordwijkerhout, The Netherlands, (pp. 477–83).

— (1986). "Mathematics Misconceptions From A Constructivist Perspective." *Annual Meeting of the American Eduational Research Association.* San Francisco.

— (1988). "Splitting and Its Relationship to Repeated Multiplication". *Tenth Annual Meeting of PME-NA.* DeKalb, IL.

— (1990). "A Review of Research on Student Conceptions in Mathematics, Science and Programming" in C. Cazden (ed.) *Review Research in Education,* **16,** American Education Research Assoc., Wash., D.C. p. 3–56.

— (1990). "Splitting, Similarity and Rate of Change: A New Approach to Multiplication and Exponential Functions." Presented at the American Educational Research Association Meeting, Boston, MA, April 16–20, 1990.

Davis, P. and R. Hersh (1982). *The Mathematical Experience.* Boston: Houghton Mifflin.

Driver, R. and J. Easley (1978). "Pupils and Paradigms: A Review of Literature Related to Concept Development in Adolescent Science Students". *Studies in Science Education,* 5, 61–84.

Driver, R. and G. Erickson (1983). "Theories-In-Action: Some Theoretical And Empirical Issues in the Study of Students' Conceptual Frameworks in Science". *Studies in Science Education,* 10, p. 37–60.

Elkana, Y. (1974). *The Discovery of Conservation of Energy*. Hutchinson University Press.
Feyerabend, P. (1978). *Science In A Free Society*. London: NLB.
Ginsburg, H. (1977). *Children's Arithmetic: How They Learn It and How You Teach It*. Austin: TX: Pro-Ed.
Goldenberg, P., W. Harvey, P. Lewis, R. Umiker, J. West and P. Zodhiates, (1988). "Mathematical, Technical, and Pedagogical Challenges in the Graphical Representation of Functions". Center for Learning Technology, Newton, MA: Education Development Center.
Goldin, G. and C.E. McClintock (1979). *Task Variables in Mathematical Problem-Solving*. Columbus, OH: ERIC Clearinghouse for Science, Mathematics, and Environmental Education, Ohio State University.
Kline, M. (1980). *Mathematics: The Loss of Certainty*. NY: Oxford University Press.
Lakatos, I. (1976). *Proofs and Refutations: The Logic of Mathematical Discovery*. J. Worrall and E. Zahar (eds.), Cambridge: Cambridge University Press.
Novak, J. and D. Gowin (1984). *Learning How To Learn*. Cambridge: Cambridge University Press.
Perkins, D.N. and R. Simmons (1987). "Patterns of Misunderstanding. An Integrative Model of Misconceptons in Science, Math, and Programming". J. Novak, *Proceedings of the Second International Seminar on Misconceptions and Educational Strategies in Science and Mathematics*. Ithaca, NY., pp. 381–95.
Piaget, J. (1971). *Biology and Knowledge*. Chicago: University of Chicago Press.
— (1970). *Genetic Epistemology*. NY: Norton and Norton.
Polya, G. (1945). *How To Solve It*. Princeton, NJ: Princeton University Press.
— (1962). *Mathematical Discovery, Vol. I & II*. NY: John Wiley & Sons.
Polanyi, M. (1958). *Personal Knowledge: Towards A Post-Critical Philosophy*. Chicago: University of Chicago Press.
Pope, M. (1985). "Constructivist Goggles: Implications for Progress in Teaching and Learning. Presented at the *BERA Conference*, Sheffield, England, 1985.
Schoenfeld, A. (1985). *Mathematical Problem-Solving*. NY: Academic Press.
Silver, E. (1982). "Thinking About Problem-Solving: Towards An Understanding of Metacognitive Aspects of Mathematical Problem-Solving". Presented at the *Conference on Thinking*, Suva Fiji, January, 1982.
Toulmin, S. (1972). *Human Understanding*. Princeton, NJ: Princeton University Press.
Trudeau, R. (1987). *The Non-Euclidean Revolution*. Boston: Birkhauser.
Tymoczko, T. (1979). "The Four-Color Problem and Its Philosophical Significance". *The Journal of Philosophy*, 76(2), 57–82.
— (1986). "Making Room for Mathematicians in the Philosophy of Mathematics". *The Mathematical Intelligencer*, 8(3), 44–50.
Unguru, S. (1976). "On the Need to Rewrite the History of Greek Mathematics". *Archive for History of Exact Sciences*, 15(2), 67–114.
Van Lehn, K. (1983). "On The Representation of Procedures in Repair Theory". H. Ginsburg (ed.), *The Development of Mathematical Thinking*. (pp. 201–53). NY: Academic Press.
von Glasersfeld, E. (1984). "An Introduction to Radical Constructivism". P. Watzlawick (ed.), *The Invented Reality*. NY: W.W. Norton.
— (1985). "Representation and Deduction." *Proceedings of the Ninth International Conference for the Psychology of Mathematics Education*. Noordwijkerhout, (pp. 484–89).

UNDERSTANDING STUDENTS' BELIEFS ABOUT PROBABILITY

CLIFFORD KONOLD

University of Massachusetts, Amherst

Probability is a particularly slippery concept. What makes it uniquely troublesome is the territory it stakes out. Through probability, we attempt to demarcate the amorphous state somewhere between the imagined extremes of total ignorance and perfect knowledge. And it is trying to keep one's footing in this nowhere land that is particularly disturbing. Like a frictionless surface, the conceptual landscape not only trips you up, but keeps you sliding once you're down.

According to Hacking (1973), prior to the 17th century such a middle ground between belief and knowledge did not exist, yet the term *probability* had already been around for some time. Hacking's thesis is that in the early to mid 17th century the concept of evidence changed such that information about the nature of things could, for the first time, be found in the things themselves. Prior to this, there existed a class of phenomena that could be known through demonstration (*scientia*) and another class that could only be testified to either by men of authority or through God-given signs (*opinio*). The word *probability* was associated with the latter class such that an opinion was probable if it had received the stamp of approval of some authority. This meaning was still around as late as the early 18th century as manifested in Daniel Defoe's 1724 novel *Roxana*, in which a woman describes her living arrangements thus:

> This was the first view I had of living comfortably indeed, and it was a very probable way, I must confess, seeing we had very good conveniences, six rooms on a floor, and three stories high.

A "probable" belief (or circumstance) was thus an "approved" one. But no amount of approval could ever make an approved belief a matter of demonstrable knowledge. Hacking argues that a bridge was created between belief and knowledge as a result of a change in, or enlargement of, the concept of evidence. The old term, *probability*, took on the function of describing the distance a belief had traveled from total ignorance to certain knowledge. At the same time, however, it also began to be used to refer to the tendency of certain phenomena (like coin flips or die rolls) to produce stable frequencies over many repetitions. This dual usage was no accident, for frequency data were a special variety of this new type of evidence that permitted degrees of belief separate from opinion. Whether a belief was probable had now less to do with approval of authority than with the approval of data.

E. von Glasersfeld (ed.), Radical Constructivism in Mathematics Education, 139–156.

If we were to meet a 16th century ancestor, we would no doubt get bogged down trying to understand what we each meant when we said that something was probable. Unfortunately, we need not time travel to observe or engage in a confusing dialogue about probability: We need only to walk into a classroom where probability is being taught to find two parties floundering in an attempt to understand one another. If Hacking is correct in his analysis, a conversation with an early ancestor would be less troublesome because of our having understood, through Hacking, their concept of evidence and what it meant to "know" versus to "believe." Similarly, I will argue in this paper that understanding how students think about probability before and during instruction can facilitate communication between the student and teacher of probability.

As other contributors to this volume stress, information cannot be picked from the external environment like apples from a tree. What a student mentally carries away from a teaching episode is not information that is received directly, but that is constructed. The construction involves the student weaving *selected* and *interpreted* teacher outputs into an existing fabric of knowledge. What a student learns from a given classroom experience is both limited and, at the same time, made possible by what he or she already *knows*.

Of course, the role of prior knowledge in learning is only half the picture. If I believed that there were no mechanism by which the external environment could impose itself in some way upon a person's conceptions, I certainly would not be writing this paper, nor would I be concerned with any other social, collective activity. I would be condemned, or delivered, depending on one's point of view, to my own internally isolated world. The nature of this "accommodating" mechanism is discussed by Piaget, von Glasersfeld, and others, and has to do with the fact that some conceptions allow me to accomplish certain goals while other conceptions do not. The fact that I cannot pick up an apple until I have learned to accommodate my hand to certain of its properties prevents me from treating the apple in a totally arbitrary way if I indeed want to grasp it. Now, of course, I'm arguing that we in fact do acquire information from the environment much like we pick apples from trees. But I hope that in the process of contradicting myself, I have alerted the reader to what might be involved in the act of apple-picking – that it involves a reciprocal relation between the apple and the picker which is more complex than we ordinarily grant.

Constructivist educators are of the opinion that the student-teacher relationship is more complex than a casual analysis would lead us to believe, and that while apple-picking can continue well enough without critical analysis, teaching cannot. While this is a debatable point, my purpose here is not to defend that opinion. I state it here as an assumption and proceed instead to demonstrate how investigations into student understandings of probability can be of pedagogical value.

Although I don't want here to defend my point of view, it will serve my purposes to exemplify what I regard as a contrary one. The following quotation is from a chapter in Cherry (1968) and is part of an introduction to Carnap's logical-relations theory of probability:

> As with all mathematics, there is no need of definition of the basic concepts, but only of the formulation of rules for using them. We do not need to know what probabilities "are," but rather how to combine them. But, nevertheless, it helps most people to have intuitive notions of the basic concepts, though such notions can become a hindrance to the expert (p. 234).

This quotation is grounded in a particular philosophy of mathematics, a theory of mathematics education, and a view of mathematical expertise. The philosophy of mathematics is one that has dominated mathematics since the 1920s. According to it, mathematics does not address questions concerning the interpretation or meaning of mathematical objects, but focuses attention instead on the validity of operations on those objects. Mathematical expertise thus involves facility with abstractions, and not necessarily with being able to apply the mathematics to some real-world problem. This philosophy has encouraged the view that to engender mathematical thinking, the teacher ought to divorce the activity in the first place from real-world experience. While it may sometimes help to provide some grounding intuitions, these are viewed as temporary supports rather than as foundation stones. Furthermore, these intuitions seem to be regarded as notions that the teacher may choose to develop.

My assumption is that students have intuitions about probability and that they can't check these in at the classroom door. The success of the teacher depends in large part on how these notions are treated in relation to those the teacher would like the student to acquire. Additionally, I think it is a myth that mathematics, either as a discipline or in the mind of a mathematician, develops independently from concerns about objects and relations that are believed to have real-world referents. This was certainly not so in the case of the development of probability theory.

I. FORMAL INTERPRETATIONS OF PROBABILITY

Our present concept of probability took form around 1660 when Pascal, Huygens, Leibniz, Fermat, and others somewhat independently developed formalizations for treating such diverse phenomena as games of chance, legal decisions and annuities. In a letter to Fermat, dated October 27, 1654, Pascal reviewed their independently-arrived-at solutions to a problem of how to divide stakes fairly in an interrupted game of chance. He demonstrated that while on the surface their approaches appeared different, they were, in fact, comparable. He concluded with the statement, "Now our harmony has begun again" (Maistrov, 1974, p. 39).

What came into being at that time was a new tune; not so new were the lyrics – "How probable is it that ...." Since that time we have been plagued (some would argue) with a concept of probability that has at least two different, but related, aspects. On the one hand it is a degree of subjective belief in the truth of some proposition; on the other, it refers in a precise way to an objective property (frequency of occurrence) of certain types of events.

There are several schools of probability, each with a somewhat different interpretation of probability. The first interpretation to which students are typically exposed is the *classical interpretation* according to which the probability of an event is simply the ratio of the number of alternatives favorable to that event to the total number of equally-likely alternatives. Thus, the probability of rolling a 3 with a fair die is 1/6, since only one alternative out of a total of six is favorable to the outcome. Introductory textbooks often refer to the classical interpretation as "theoretical probability." This interpretation is obviously limited to trials with objects such as coins, dice, and spinners, that are composed of equally-likely alternatives. It is also logically flawed in that its definition of probability is circular: Probability is defined in terms of equally-likely alternatives, yet what can be meant by "equally-likely" other than "equally-probable?"

Perhaps the most pervasive objective theory of probability is the *frequentist interpretation*, which defines probability in terms of the limiting relative frequency of occurrence of an event in an infinite, or near infinite, number of trials. Thus the probability of rolling a 3 with a particular die is the ratio of trials resulting in a 3 to the total number of trials (actually the limit of this ratio as the number of trials approaches infinity). In introductory textbooks, the frequency interpretation is sometimes referred to as "empirical probability." While this interpretation can be applied to events that are composed of non-equally-likely alternatives, it is restricted to setups such as coin tossing and urn drawings with which "identical" trials can be repeated indefinitely. In addition, since probability is defined as frequency in the long run, it is meaningless from this perspective to talk about the probability of an event occurring on a particular trial. It should be pointed out that even though this interpretation is referred to as an objective theory, this does not mean that it is free from subjective considerations. Most importantly, this interpretation requires an observer who can count a series of events, and in order to accumulate a sum of occurrences, that observer must consider the events which are counted to be of the "same type." If each flip of a particular coin is considered unique in an essential way, and there are certainly grounds for doing so, then the frequency interpretation cannot be applied.

One advantage of various *subjectivist interpretations* of probability is that they can be applied to a wide range of phenomena since, according to this view, probability is a measure of belief in the truth of a proposition. According to a subjectivist view, different people could validly assign different values to the probability of rolling a 3 with a particular die. These values would presumably reflect different beliefs about the fairness of the die, the character of the person doing the rolling, the technique used in rolling, etc. However, in formalizing subjectivist interpretations, theorists have adopted various adjustment mechanisms (e.g., Bayes' theorem) that lead to the revision of initial probabilities given new information such as results of actual trials. Given enough data about the frequency of occurrence of 3 with a particular die, the various subjective probabilities held by different people would all begin to converge on the frequentists' limit.

The meaning of the probability value in a subjectivist interpretation can be thought of in several ways. One of the most common is to describe the value as a measure of a person's belief in what would constitute a fair bet. Thus, a person estimating the chance of rain at 70% would as quickly bet $7 to win $10 if it did rain as they would bet $3 to win $10 if it didn't rain. Another way to think about the meaning of the value is to consider as a collection all those events to which a person would assign the probability 70%. We could then look to see how many of these predicted events actually occurred. If only 50% of the events which had been assigned 70% probability actually occurred, then we would conclude that this person was, in general, overconfident in his or her predictions. If 90% of the events actually occurred, then the person would be underconfident. The nearer the percentage of predicted events came to 70%, the more expert we would regard the person in arriving at subjective probability judgments. In the lingo of decision theory such a person would be "well-calibrated." Thus, if we assume that a person is well-calibrated, when this person assigns the probability p% to events we should expect roughly p% of those events to occur.

Notice that the subjectivist theories are *normative* theories, specifying how "rational" people ought to formulate and alter their beliefs in the light of new information; they are not *descriptive* theories of how people actually formulate and alter subjective probabilities. I use the term "normative" in this chapter to refer to theories or beliefs that are held by those regarded as experts. When I speak of "normative theories of probability," I will be referring not to any particular theory of probability but collectively to all those theories that are considered by experts in the field as deserving serious attention. Thus, the term "normative" is not synonymous with "correct," but is very similar to the pre-modern meaning of probable, i.e., "approved."

These various interpretations are mentioned here primarily to provide a context for evaluating various beliefs about probability held by students. I want to proceed by suggesting that two elements of Hacking's account of the development of the modern conception of probability have particularly important implications for the teaching of probability. First, probability is viewed by Hacking as a concept that exists in a web of discourse and related concepts. Thus, the modern concepts of frequency of occurrence could not *emerge* until the related concept of evidence changed. Second, once a new, objectified notion of probability had appeared, it wasn't totally divorced from its prior, epistemological meaning. Hacking claims that concepts "play out their own lives in, as it were, a space of their own" (p. 15). Concepts are not freely created and revised at will. Rather, they are subject to restrictions that are inherent in the web of concept-relations in which they are embedded. The meaning of a word cannot be directly enforced; it emerges from and is supported by a tenuous consensus among those who use the word. Thus, despite efforts of advocates of both objective and subjective schools of thought to annex the concept of probability, it maintains even today its Janus face.

I want to argue that these two features also apply to communications between students and teachers regarding probability. The teacher has a particular concept of probability and intends to communicate, in some way, this concept to the students. Ideally, the teacher's view is similar to one of the interpretations summarized above. To the extent that the conceptual frameworks of the teacher and the student are compatible, a sense of mutual understanding is possible. Where those frameworks are incompatible, however, communication becomes problematic. The teacher cannot, by decree, enforce a normative view. The primary reason for this inability to simply transfer a concept from teacher to student is that students have no other option than to *interpret* what the teacher says or does in light of what they already understand of probability. As Fisher and Lipson (1985) have written:

> ... what we perceive is strongly influenced by our beliefs and expectations. The constructive nature of thought has the consequence that every observation is an interference and is dependent upon our existing knowledge of the world. In spite of this, we frequently become aware of differences between our expectations and observations. As a result we are more or less constantly engaged in assessing the "goodness-of-fit" between our mental models and our experience with the world around us (p. 50).

Not only does this quotation clearly express the idea that what we perceive and experience as existing in the external world is dependent on our prior knowledge, it also inadvertently demonstrates the very point. The word *interference* in the second sentence ought to read *inference*. Whether or not a reader notices and resolves this error depends on prior knowledge. Some readers, like the authors of the article, who are very familiar with the expressed idea, are likely to see what they expect to see and read the word as *inference*. Others read the word as *interference*, but knowing the point that the authors are making can fairly quickly replace it with the intended word. In this way they validated the claim made in the last part of the quote. Those totally unfamiliar with the authors' view of perception are more likely to read the word as *interference* and perhaps create an interpretation with which this word fits.

Long before their formal introduction to probability, students have dealt with countless situations involving uncertainty and have learned to use, in common discourse, words such as *probable, random, independent, lucky, chance, fair, unlikely*. They have an understanding that permits them to use these words in sentences that are comprehensible to other language users in everyday situations. It is into this web of meanings that students attempt to integrate and thus make sense of their classroom experience. Unfortunately, since much of formal probability theory does not fit well into the informal understandings that students already possess, misunderstandings, serious confusions, and other breakdowns in communication occur. If a teacher can come to see how students view probability, that teacher will be better able, emotionally as well as conceptually, to aid the development of a normative concept. With this goal in mind, I describe in the next section some of the non-normative beliefs and perceptions about probability that have been found to be commonly held.

II. STUDENT CONCEPTIONS

The view that statistics and probability are conceptually difficult topics for the untutored is relatively recent. Up through the 1960s it was generally assumed that when people reasoned in everyday situations about uncertainty they used methods similar to those a statistician would use, but less carefully (Peterson & Beach, 1967). Piaget and Inhelder (1951) claimed that by the age of 12 most children could reason probabilistically about a variety of randomizing devices and had developed sound statistical notions including an intuitive understanding of the Law of Large Numbers.

More recently, cognitive psychologists have found pervasive and persistent errors in peoples' reasoning under uncertainty. Contrary to the earlier assumptions, research indicates that people arrive at probabilistic judgments via considerations that are qualitatively different from the statisticians'. Daniel Kahneman and Amos Tversky (1973) have offered the most comprehensive account to date of why peoples' judgments often differ from accepted theory. They suggest that because of limited information-processing capabilities, people use various judgment heuristics that allow them to summarize large amounts of data and quickly arrive at decisions.

Although these heuristics usually produce adequate estimates, they are limited in the amount and type of information to which they are sensitive. This, in turn, leads to predictable judgment errors in some situations. For example, most people incorrectly believe that the ordered sequence MMMMMM of male and female births in a family is less likely than the ordered sequence MFFMMF. One possible explanation of this belief is that people confuse the latter sequence with the unordered outcome "3 Ms, 3Fs." If the precise order of births is disregarded, it is certainly the case that the outcome "3 Ms, 3Fs" is more likely than "6 Ms." (There are 20 different ways to arrange 3 Ms and 3 Fs and only 1 way to arrange 6 Ms.) Kahneman and Tversky (1972) suggest, however, that MFFMMF is judged as more likely through the application of what they term the "representativeness heuristic." According to this heuristic, the probability of a sample is estimated by noting the degree of similarity between the sample and parent population. Since the sequence MFFMMF is more similar to the population proportion of approximately half males and half females, and also appears to better reflect the random process underlying sex determination, it is judged as more likely.

The heuristics described by Kahneman and Tversky are useful in explaining why so many results of probability theory seem counter-intuitive and incorrect to the student. In fact, it has been demonstrated that even experts in probability can be led to the unconscious application of heuristics for situations that they know call for a probabilistic analysis (Tversky & Kahneman, 1971). Heuristics thus appear to be deeply held and, in many cases, automatically applied. They have been compared to optical illusions where, even though one may know better, the situation cannot be perceived veridically.

Another way to describe these heuristics is that they provide an account of how people with no formal training derive an estimate of the probability of an event. Hidden in the heuristic account is the assumption that regardless of whether one uses heuristics or the formal methods of probability theory, the individual perceives the goal as arriving at the probability of the event in question. While the derived probability value may be non-normative, the meaning of that probability is assumed to lie somewhere in the range of acceptable interpretation.

I have been studying a pattern of errors that are more fundamental, in some respects, than those resulting from the use of judgment heuristics. Rather then exploring how people *arrive* at probabilistic judgments, I have been interested in how people *interpret* a question about probability or a probability value. Results of interviews in which college students reasoned aloud about situations involving uncertainty indicate that for many students a question about probability is interpreted in a way that is, from a normative view, non-probabilistic (Konold, 1989).

According to this alternative interpretation, which I will refer to as the "outcome approach," the primary goal in situations involving uncertainty is not to arrive at a probability of occurrence but to successfully *predict* the outcome of a *single* trial. Given this objective, a question that explicitly asks for the probability of an outcome is interpreted as asking whether the outcome will, in fact, occur on the next trial. For example, asked to explain a weather forecaster's prediction of 70% chance of rain, many students respond that they take that to mean that it *will* rain. Asked what they would conclude if it did not rain, these same students hold that the forecaster's prediction would then have been wrong. They also will argue that a forecaster is performing sub-optimally when it rains on 70% of the days for which 70% chances were given. Probability values are evaluated in the outcome approach in terms of their proximity to the anchor values of 100%, 0%, and 50%, which have the respective meanings of "yes," "no," and "I don't know." Thus students reasoning according to the outcome approach will argue that 70% is sufficiently close to 100% to warrant the assertion, "It will rain tomorrow."

Students who show this preference for predicting individual as opposed to sample outcomes also tend to rely on causal as opposed to chance explanations of outcome occurrence and variability. For example, some students believe that the number in the estimate "70% chance of rain" refers to a measure of some factor that causes rain. Thus they believe that the 70% implies 70% humidity, or

70% cloud cover. While predictions are often based on inferences about causality, frequency information is often ignored. For example, I gave students an irregularly-shaped bone to roll and asked them which of the six surfaces was most likely to land upright. As mentioned above, many of the students appeared to interpret the question as a request to predict the outcome of a single trial and accordingly evaluated their predictions as being correct or incorrect after one roll of the bone. And, remarkably, after conducting several trials with the bone and being shown a summary of results of 1,000 trials, some students preferred to base their predictions on a visual inspection of the bone rather than on the available data.

In the remainder of this section I will present and discuss in some detail a few excerpts from conversations with students about probability. My purpose is to demonstrate how statements that otherwise would be regarded as contradictory or incomprehensible can be understood if one assumes that students are reasoning according to the outcome approach. More generally, I want to suggest that understanding how students are thinking about a topic – in this case probability – puts the teacher in the position of being able to initiate conversations and design curriculum that can facilitate the development of more adequate concepts of probability. Without this type of careful attention to student thinking, instruction in probability will, for many students, have no lasting impact.

2.1 Making Confusing Statements Understandable

The excerpt presented below demonstrates how statements that would otherwise seem disconnected and incomprehensible can be seen, with the aid of a model of student reasoning, as both comprehensible and connected. The conversation took place in an experimental setting in which the student had been asked what a 70% chance of rain meant, and is here responding to the follow-up question:

I: Suppose you wanted to find out how good a particular forecaster's predictions were. You observed what happened on ten days for which a 70% chance of rain had been reported. On three of those ten days there was no rain. What would you conclude about the accuracy of this forecaster?

The student first responded to this probe by concluding that the forecaster was "very accurate" but then began to question whether, in this instance, one could look beyond the prediction of an individual day in assessing accuracy:

S: I was just wondering whether – if you're taking a ten-day span and, as you said, three of the days it didn't rain, if that can really relate to when he's looking at an individual day – that particular day. And I suppose it can if you're looking at a ten-day span with 70% chance of rain every day, with the same setup.

When asked if the forecaster could have been any more accurate she apparently encoded the question as, "Should the forecaster have predicted higher than 70% chance?" She responded that giving higher proportions does not imply higher accuracy:

S: No, I don't think so. No 'cause when he says there's going to be 70% chance of rain, he really can't be more – I don't think he can be more predictive than that. 'Cause that's a proportion. If he said 50% chance of – that's, you know, not any more accurate than if he said 70% chance of rain.
I: Saying 50% chance of rain is no more
S: Yeah.
I: accurate? In terms of what?
S: I guess, as I said, he must have certain standards to go by when he picks a chance of rain.

In an attempt to get her to clarify her statements, she was further probed:

I: Say the forecaster predicted 50% chance of rain on those ten days. What would have happened if the forecaster was really good? If you kept track, what would you expect in terms of the number of days it would rain and the number of days it wouldn't rain?
S: Actually, you couldn't really expect anything. Because he is looking at an individual day, 50 and 50%. So let's say if it rained that day, then he had a – he was – I don't know, 'cause you're looking at individual days, really, so it could have rained all the time, it could have rained not one day out of ten days, and then it could have been 50/50, like five days it rained, five days it didn't rain, and he wouldn't be – and it would be the same, actually. It would come out the same, I guess, 'cause he is looking at individual days.
I: Tell me again what would come out the same. If over ten days it didn't rain at all –
S: Yeah, and if it did rain. 'Cause he's looking at a particular day. And it's 50% chance rain, 50% not. So he wouldn't be more or less accurate in any of those situations, I don't think.

These rather confusing statements become fairly comprehensible when the student is viewed as interpreting the 50% in a way that is consistent with the outcome approach. According to the outcome approach, 50% is the mid point of the yes/no decision continuum, and thus it means that anything can happen – "It may rain, or it may not rain. Who knows?" Under this interpretation the 50% carries no implication concerning the frequency of rain. As a result of considering the meaning of a 50% forecast, this student confirms her conclusion that assessing accuracy over days is inappropriate. She is apparently aware of the implications involved in assessing the accuracy of a forecast of 70% chance of rain by looking at the frequency of rain on days for which a 70% chance of rain is given. To be consistent with this, one would expect a 50% forecast for 10 days to be accurate if it rained on 5 of those days. But this interpretation of accuracy would contradict the use of 50% as an indicator of "anything-could-happen" since according to the latter interpretation, any imaginable result of a ten-day sample would seem equally consistent with a forecast of 50% rain. This reasoning makes her more certain that individual days is the appropriate unit of analysis, and accordingly, when asked to summarize what she believed, she responded:

S: Well, he's looking at an individual day – particular day – and he's setting up percentages on one day. And you can't really extend that to an amount of time.

There is a tendency for teachers, when confronted with an apparently incorrect statement, to inform the student of the error and perhaps state the correct point of view. For example, one can imagine at some point in the above discussion telling the student, "No, 50% would mean that the forecaster would expect rain on 5 out of 10 such days, just as 70% means that rain would be expected on 7 out of 10 such days." However, if a student's statements do not reflect isolated beliefs but are grounded in a more general conceptual framework, then one must question the effectiveness of these types of local interventions; they do not get at the heart of the problem. An analogous situation would involve correcting the assertion that a ship would fall off the earth if it ventured too far from shore by negating the assertion or citing evidence to the contrary rather than focusing on the apparent underlying belief in a flat rather than spherical planet. By understanding the outcome approach and the nature of various judgment heuristics, the teacher can "see beneath" superficial features of various student claims.

2.2 Statements That Sound Incorrect

The following is an example of a claim that seems patently incorrect. The conversation with a high school student who was participating in a summer mathematics program at Mount Holyoke College occurred spontaneously about half way through a two-week workshop on probability:

S: I don't believe in probability.
I: Why?
S: Because even if there is a 20% chance, it could happen.
I: Yes.
S: Even 1%, it could happen. I don't believe in probability.

On first reading, this student's claim seems baffling. If the assumption is made, however, that this student uses the term *probability* in a way consistent with the outcome approach, then it ought to be given the approximate meaning, "predicting single trials." Indeed, if that phrase is substituted for the word *probability*, the claim becomes not only comprehensible but normative. My interpretation is that this student is beginning to question the validity of her prior outcome-oriented view of probability. She is saying that she no longer believes that small probability values can be regarded as zero, and therefore one cannot predict with certainty the outcome of a chance event. She isn't yet aware, however, that the word *probability* can refer to something other than single-trial predictions.

It should be pointed out that the outcome approach is reasonably consistent with one of the common meanings of the related word *probable*. A probable event is, in everyday usage, one that is likely to occur. For example, George Shultz was reported in *The New York Times* (October 1, 1987) as saying, "I think there's not just a possibility, I think there's a probability," in response to a question about the likelihood of an eventual arms embargo against Iran by the Soviet Union. When asked to define probability, students often provide definitions of the type, "Probability is the chance of one thing occurring instead of another," or "Probability is what is most likely to be." This usage of the word *probability* to imply that some event is more likely to occur than not, fits well with students' outcome-oriented belief that mathematical probability involves *deciding whether* an event will occur, rather than *quantifying how often* that event will occur.

2.3 Two Views in Conflict

Although students can often fit what they read or hear in the classroom into their preconceived notions of probability, conflicts frequently arise between their view and the view being advocated by the teacher. These conflicts are welcome by the constructivist teacher. In the attempt to resolve the conflict, students often alter their understanding, bringing them into closer agreement with the normative perspective. An awareness of the student's beliefs permits the teacher to understand the nature of the conflict and possibly to help the student to navigate successfully. In the first excerpt discussed above, the student was experiencing some conflict over the appropriate unit of analysis in deciding the accuracy of weather forecasts. She resolved the conflict not by modifying her notion of probability but by reaffirming her belief that individual predictions are the appropriate unit. The excerpt below involves a student who experienced a similar conflict in responding to the probe about the accuracy of weather forecasts. He could not, during this conversation, come to a resolution, but moved back and forth between viewpoints in what he called a "logic swirl."

S: Oh, it seems as though they were pretty much right on because in the whole of the ten days they were right on three of the occasions, which would be 30% it didn't rain, and that's what they kind of predicted, almost. Maybe.

I: What are you thinking?

S: Well, they didn't predict a ten-day span. They predicted each individual day. And so it seems as though when each day comes up, it's a whole in itself, and it's not necessarily put together in a unit.

S: ... It kind of gets back to the idea that they were pretty much right getting seven out of ten right. Maybe it's not fair to judge it that way. Maybe you should just judge each one. But I guess you can add them together because they're all the same. They're all like 70%.

I: ... Would they have been more accurate had it rained on all ten of the days? Would that be more impressive to you?

S: Well, that's weird cause it almost seems that ... they're going over and they're wrong the other way. It's raining more than they really predicted. But it's not like they predicted that it would rain on 70% of the next ten days. It's like they predicted rain for each day. And if they were 70% sure ... that it was going to rain on each day ... if it rained on more of those days, then as you increase the number above seven, then they would be less accurate.

The interaction described below took place during a lab on probability that was being run in conjunction with a remedial-level mathematics course at the University of Massachusetts. This particular episode is interesting both because of the extreme attempt made by the student to maintain his single-trial interpretation and also because by the end of the semester he had formulated a normative interpretation. He referred to this episode as being particularly important in the development of his understanding of probability.

The lab activity involved students guessing at the probabilities associated with a thumb tack landing with the point up (U) or the point down (D). For most thumb tacks, the probability of U is a bit greater than D. After making a guess as to the probabilities, each student conducted and recorded the results of 100 trials with the tack and then used these results to evaluate and, if necessary, revise the probability estimates. This particular student had initially guessed that the probability of U was .70. He then conducted 100 trials, and in spite of the fact that he had obtained 55 Us and 45 Ds, he concluded that D was more likely. I asked:

I: Suppose you had to bet on whether you'd get more Ds or Us in another 100 trials. What would you bet on?
S: I'd bet on D.
I: Why?
S: Because I got more Ds when I dropped it.
I: According to your results you got 55 Us and 45 Ds.
S: [Long pause while he verifies this result.] Well, ya. But I think D is more likely.
I: Can you explain why you think that, even though you got more Us?
S: Because it seemed like D came up more often when I dropped it once.
I: Which would be more likely if you dropped it 100 times?
S: OK. Well I guess U is more likely if I do it 100 times, but if I did it just once, D would be more likely.

This student eventually saw the inconsistency in his reasoning, but his last effort to salvage his belief that D was more probable by adopting opposite predictions for a single trial vs. 100 trials is a dramatic demonstration of the preference for thinking about probability in terms of single trials.

After resolving this conflict, the student became curious about why, having begun thinking U was more likely, he ended up thinking D was more likely even though he got more Us. He finally noticed that the first few trials were predominantly Ds and remembered that this had surprised him. He reasoned that thereafter he had "paid more attention" to trials resulting in D than those resulting in U. Forgetting for the moment whether his self-analysis was accurate, we see that his explanation is quite sophisticated and is consistent with an effect that Tversky and Kahneman (1973) have described as resulting from the "availability heuristic." According to this heuristic, the probability of an event is estimated by assessing the relative ease of bringing that event to mind. Thus people may determine the probability of winning a lottery by trying to recall people they know, or know of, who have won. Presumably, this is one reason that state lotteries and sweepstakes like to advertise using the names and homey photos of past winners – to provide us with a set of instances of winners whom we sort of know or perhaps can imagine knowing. One of the factors that can influence the ease of recalling event-instances is the vividness or saliency of an event. Witnessing a horrible traffic accident can, for a while, elevate a person's subjective sense of the probability of being in an accident. This student's account of why he came to think that the thumb tack was more likely to land with the point down appears consistent with the above account. Seeing what he didn't expect led him to "pay more attention" to subsequent occurrences of the same type such that his memory of the frequency of that event was distorted.

III. A FITTING APPROACH TO INSTRUCTION

I have suggested that students come into their first course on probability with a reasonably coherent and deeply engrained point of view about probability. One component of students' prior knowledge about probability involves judgment heuristics that operate much like visual perceptions. Thus, HTTHT just *looks* to most people to be a more likely result of flipping a fair coin than HHHHH. In addition to these perceptions, students also have a different conceptual understanding of the goal in situations involving uncertainty, i.e., that the goal is to predict individual outcomes. Together, these cognitions make learning the normative view difficult.

Although these informal models and methods pose a serious challenge to those trying to teach and learn probability, there is emerging evidence that instruction in misconception-rich domains can be effective. In physics, where students also have strong prior conceptions, it has been suggested that it may be effective to encourage students to recognize and resolve conflicts between normative concepts and erroneous intuitions (McDermott, 1984). Several researchers, including Minstrell (1982), Clement (1987), and Hake (1987), have demonstrated that physics instruction specifically designed to address various misconceptions can be effective. Their approach includes laboratory exercises designed to demonstrate counter-intuitive results and promote student discussion, problems that require qualitative rather than quantitative solutions, as well as presentations that explicitly contrast normative with non-normative physics concepts.

Inspection of the various classroom techniques being used to overcome physics misconceptions reveals three general criteria against which students are encouraged to evaluate their current beliefs: the fit between their beliefs and 1) the beliefs of others, 2) their other, related beliefs, and 3) their own observations. I conclude this chapter with a brief discussion of these criteria with the conviction that if instruction in probability is to be effective, the teacher will need not only to be attuned to student intuitions but to structure activities that encourage students to evaluate those intuitions in accord with these three criteria.

1. *Do my beliefs agree or fit with the beliefs of others?* A typical starting point for the presentation of some physics concept or principle is to invite students to discuss among themselves what they believe about a particular situation. For

example, Minstrell (1982) asks students to determine the forces that are acting on a book lying on a table. After allowing some time for students individually to think about the question and to draw a diagram showing all of the forces acting on the book, he has students share their answers with the class. This sharing leads to a dialogue, with students questioning and challenging one another. It should be pointed out that discussions that allow for the expression and defense of differing points of view are not common in the everyday experiences of the students. They are certainly not the norm in the classroom nor do they occur in typical social situations. As Stubbs (1983) observes,

> ... there is a general rule in our society that demands that interaction proceed at a smooth flow: silences are often considered embarrassing and disagreements must normally be mitigated. So speakers immediately counteract departures from the smooth ongoing of normal face-to-face interaction by making (if necessary, violent) attempts to restore the ritual equilibrium (p. 241).

Thus, even though instructors may say little during this discussion, they play a critical role in keeping the conversation going, being supportive of students' expression of their views, and at times helping to focus the discussion, but above all, creating an atmosphere in which it is acceptable to articulate an opposing view and to challenge what one does not understand or believe.

The class discussion accomplishes a number of goals. First, it allows students to make explicit to themselves what they believe about a situation before they know what the "expert" view is. Second, classroom discussion provides a motivation for further exploration of the question. The act of articulating one's beliefs in a public forum involves a personal investment in the question. Students who express their view are no longer indifferent to the final outcome of the discussion – in general they want to have been "right" rather than "wrong." Third, discussion among students provides the opportunity for the instructor to gain further insights into how students are thinking about a particular topic. This information can be used to plan future interventions and to monitor conceptual development. finally, the discussion communicates to the student the value that the instructor places on students' understanding and expression of ideas.

Once students have made their own beliefs more explicit, the second criterion can be applied:

2. *Are my beliefs internally consistent?* In the process of trying to convince one another, students explore in greater depth the implications of and interconnections among their own beliefs. At times this exploration can lead students to discover inconsistencies in their beliefs or gaps in their reasoning. For example, Minstrell (1982) asks a series of questions related to the book-on-the-table problem mentioned above. Roughly the same number of students who initially do not believe that a table exerts an upward force on a book resting on the table, do not believe that an upward force is exerted on a book resting on an out-

stretched hand. However, as more books are added to the outstretched hand, more students subsequently express a belief in an upward force, and the majority of students believe that there is an upward force on a book resting on a spring. As the students progress through these questions, they are repeatedly asked about the book on the table, and an increasing number indicate a belief that the table pushes up on the book. The students are apparently motivated to see the situations as analogous and therefore to be consistent.

In addition to asking students to extend their analysis to a variety of related instances, various probes can be effective in encouraging students to analyze the consistency and completeness of their beliefs. These include probes of the form,

Why do you think that ...?
How would you explain a situation where ...?
How does that relate to what you said earlier?
What would you say to a person who gave the following argument ...?

One reason that misconceptions are so difficult to alter is that they tend to comprise a coherent, self-consistent framework. To the extent that a person's beliefs are self-consistent, they are impervious to this type of challenge. This is one reason why the third criterion is especially important.

3. *Do my beliefs fit with empirical observations?* Beliefs about physical, social and probabilistic phenomena are always to some extent based on or confirmed by observations of events. This includes beliefs that are regarded by the "educated" as irrational: People who believe in astrology will argue that predictions based on their astrological chart have come true, and paranoid schizophrenics will cite numerous observations to support their claim that others are conspiring against them. The fact that what one observes is never independent of what one already knows or believes, means that data cannot serve as the final arbiter of so-called "empirical questions." How is it then that the fit between empirical observations and beliefs can be considered a criterion in evaluating those beliefs? This is a complex question which I cannot discuss adequately here. But there are two things that people generally do not do that further weakens the informative power of empirical observations: 1) they do not keep accurate records and 2) they do not look for data that would be inconsistent with and thus controvert, a belief they hold. To be effective, classroom demonstrations meant to challenge students' beliefs should emphasize these two features.

Although demonstrations have been a traditional part of physics instruction, they seldom involve students first speculating and making predictions of what they think will happen based on their own understanding of the situation. Minstrell (1982; 1984) has used demonstrations along with this type of discussion and prediction very effectively. With the arrival of the computer in the classroom, similar demonstrations of probabilistic and statistical phenomena are now practicable. Large random samples can be quickly and repeatedly drawn and summary results computed and displayed. By having students first make predictions of what they expect to observe, such simulations can be used to challenge non-normative beliefs about random events.

The three criteria for evaluating current beliefs are appropriate for use in the constructivist classroom not only because they are effective in inducing conceptual change but also because they are the same criteria that define scientific activity. Scientific theories that are worth further consideration must first be made explicit, they must not contain internal contradictions, and they must fit with observations. Students with strong prior conceptions are therefore treated in this instructional approach as scientists, in as much as they are viewed as holding theories which they should not be expected to abandon until these theories have proven unsuccessful (and not just unsuccessful vis-à-vis course grades). Furthermore, the testing of the adequacy of student beliefs should not be an isolated, individual enterprise but, as in scientific activity, involve a dialogue that takes place among a community which negotiates the relative merit of various perspectives.

REFERENCES

Cherry, C. (1968). *On human communication.* Cambridge, MA: M.I.T. Press.

Clement, J. (1987). Overcoming students' misconceptions in physics: The role of anchoring intuitions and analogical validity. In J. D. Novak (Ed.), *Proceedings of the Second International Seminar, Misconceptions and Educational Strategies in Science and Mathematics* (pp. 84–97). Ithaca, NY: Cornell University.

Fisher, K. M., & Lipson J. I. (1985). Information processing interpretation of errors in college science learning. *Instructional Science, 14*, 49–74.

Hacking, I. (1975). *The emergence of probability.* London: Cambridge University Press.

Hake, R. R. (1987). Promoting student crossover to the Newtonian world. *American Journal of Physics, 55*, 878–884.

Kahneman, D., & Tversky, A. (1972). On prediction and judgment. *Oregon Institute Bulletin, 12*(4).

— (1973). On the psychology of prediction. *Psychological Review, 80*, 237–251.

Konold, C. (1989). Informal conceptions of probability. *Cognition and Instruction, 6*, 59–98.

Maistrov, L. E. (1974). *Probability theory.* New York: Academic Press.

McDermott, L. C. (1984). Research on conceptual understanding in mechanics. *Physics Today, 37*, 24–32.

Minstrell, J. (1982). Explaining the "at rest" condition of an object. *The Physics Teacher, 20*, 10–14.

— (1984). Teaching for the understanding of ideas: Forces on moving objects. In C. W. Anderson (Ed.), *Observing science classrooms: Perspectives from research and practice* (A.E.T.S. Yearbook XI). Columbus, OH: ERIC Center for Science, Mathematics and Environmental Education.

Peterson, C. R., & Beach, L. R. (1967). Man as an intuitive statistician. *Psychological Bulletin, 68*, 29–46.

Piaget, J., & Inhelder, B. (1975). *The origin of the idea of chance.* New York: Norton. (Original work published 1951)

Stubbs, M. (1983). *Discourse analysis: The sociolinguistic analysis of natural language.*

Chicago: University of Chicago Press.
Tversky, A., & Kahneman, D. (1971). Belief in the law of small numbers. *Psychological Review, 76*, 105–110.
— (1973). Availability: A heuristic for judging frequency and probability. *Cognitive Psychology, 5*, 207–232.

A CONSTRUCTIVIST APPROACH TO SECOND GRADE MATHEMATICS

PAUL COBB, TERRY WOOD, AND ERNA YACKEL

Purdue University

Our overall objective in this paper is to share a few observations made and insights gained while conducting a recently completed teaching experiment. The experiment had a strong pragmatic emphasis in that we were responsible for the mathematics instruction of a second grade class (7 year-olds) for the entire school year. Thus, we had to accommodate a variety of institutionalized constraints. As an example, we agreed to address all of the school corporation's objectives for second grade mathematics instruction. In addition, we were well aware that the school corporation administrators evaluated the project primarily in terms of mean gains on standardized achievement tests. Further, we had to be sensitive to parents' concerns, particularly as their children's participation in the project was entirely voluntary. Not surprising, these constraints profoundly influenced the ways in which we attempted to translate constructivism as a theory of knowing into practice. We were fortunate in that the classroom teacher, who had taught second grade mathematics "straight by the book" for the previous sixteen years, was a member of the project staff. Her practical wisdom and insights proved to be invaluable.

It appears that we have had some success in satisfying the institutional constraints. The achievement test scores did rise satisfactorily, the parents were all universally supportive by the middle of the school year, and the administrators developed a positive opinion of what they saw. As a consequence, we are currently working with 18 teachers from the same school system.

In general, we hope that our on-going work constitutes the beginnings of a response to Brophy's (1986) challenge that "to demonstrate the relevance and practical value of this point of view for improving school mathematics instruction, they [constructivists] will need to undertake programmatic development and research – the development of specific instructional guidelines (and materials if necessary) for accomplishing specific instructional objectives in typical classroom settings" (p. 366). Thus, we concur with Carpenter's (1983) observation that "If we are unable or unwilling to provide more direction for instruction, we are in danger of conceding the curriculum to those whose basic epistemology allows them to be more directive" (p. 109). Constructivism as an epistemology is, for us, a general way of interpreting and making sense of a variety of phenomena. It constitutes a framework within which to address situations of complexity, uniqueness, and uncertainty that Schon (1985) calls

E. von Glasersfeld (ed.), Radical Constructivism in Mathematics Education, 157–176.

"messes," and to transform them into potentially solvable problems. Thus, like any epistemology, constructivism influences both the questions posed and the criteria for what counts as an adequate solution. Its value to mathematics education will, in the long run, depend on whether this way of sense making, of problem posing and solving, contributes to the improvement of mathematics teaching and learning in typical classrooms with characteristic teachers. If it eventually fails to do so, then it will become irrelevant to mathematics educators.

I. OVERVIEW OF THE PROJECT

1.1 Rationale for Instructional Activities

Other contributors to this volume present constructivist analyses of the process of learning mathematics. It should be clear that, for the constructivist, substantive mathematical learning is a problem solving process (Cobb, 1986; Confrey, 1987; Thompson, 1985; von Glasersfeld, 1983). In this context, substantive learning refers to cognitive restructuring as opposed to accretion or tuning (Rumelhart & Norman, 1981). Consequently, our primary focus, as we developed, implemented, and refined instructional activities, was on that aspect of cognitive development that is both the most significant and the most difficult to explain and influence.

At the risk of over-simplification, an immediate implication of constructivism is that mathematics, including the so-called basics such as arithmetical computation, should be taught through problem solving. This does not mean that the instructional activities necessarily emphasize what are traditionally considered to be problems – stereotypical textbook word problems. In fact, the general notion that problems can be given ready-made to students is highly questionable. Instead, teaching through problem solving acknowledges that problems arise for students as they attempt to achieve *their* goals in the classroom. The approach respects that students are the best judges of what they find problematic and encourages them to construct solutions that they find acceptable given their current ways of knowing. The situations that children find problematic take a variety of forms and can include resolving obstacles or contradictions that arise when they use their current concepts and procedures, accounting for a surprising outcome (particularly when two alternative procedures lead to the same result), verbalizing their mathematical thinking, explaining or justifying a solution, resolving conflicting points of view, and constructing a consensual domain in which to talk about mathematics with others. As these examples make clear, genuine mathematical problems can arise from classroom social interactions as well as from solo attempts to complete the instructional activities.

In general, the instructional activities, classroom organization, and flow of the lessons were designed to facilitate the occurrence of mathematical problems that the children could attempt to resolve in conducive settings. Detailed models of early number learning (Steffe et. al., 1983; Steffe, Cobb, and von Glasersfeld, 1988) proved particularly relevant to the development of the activities. The models' specification of both young children's meaning-making capabilities and the sensory-motor and conceptual activities upon which they can reflect guided the analysis of the problematic situations that might arise for children at different conceptual levels. The instructional activities are appropriate to the extent that they engender reflective, problem solving activity on the part of the children.

1.2 Classroom Organization

The instructional activities are of two general types: Teacher-directed whole class activities and small group activities. To the extent that any lesson can be considered typical (Erickson, 1985), the teacher might first spend at most five minutes introducing the small group activities to the children. Her rationale for doing so is to clarify the intent of the activities. She might, for example, ask the children what they think a particular symbol means or ask them how they interpret the first activity. In doing so, she does not attempt to steer the children towards an official solution method but instead tries to ensure that the children's understanding of what they are to do is compatible with the activity as intended by its developers. Any suggested interpretation or solution, however immature, is acceptable provided it indicates the child has made the appropriate suppositions.

Once the teacher is satisfied that conventions used to present the instructional activities have been established, she tells them to start work. One child hands out an activity sheet to each group of two or, occasionally, three children. As the children work in groups for perhaps 25 minutes, one notices that the teacher spends the entire time moving from one group to the next, observing and frequently interacting with them as they engage in mathematical activity. In this setting she is free of managerial concerns and can focus attention on children's thinking and interactions. Because the children are encouraged to take responsibility for their own learning and behavior in the classroom, she is rarely interrupted by a child from another group asking for help or for permission to, say, use particular manipulative materials. Further, the teacher does not feel the need to monitor the children's conduct – she is comfortable giving her attention entirely to the group she is currently observing.

The noise level is somewhat higher than most classrooms because the children are talking about the activities with their partners. However, the teacher does not have to remind the children to lower their voices because the noise level always stays within reasonable bounds and she has learned that their talking is generally about mathematics and is not distracting to others. One soon notices the purposeful way that children move around the classroom on their own initiative. Some go to a table to take one of the available manipulatives that they have decided is needed. Others get additional activity sheets or perhaps a piece of scrap paper. It soon becomes apparent that some groups have completed four or five activity sheets while others might complete just one, and then only with the teacher's assistance. This disparity in the number of activities completed seems, for the most part, to be inconsequential to both the teacher and the children. Finally, the teacher tells the children when there is only one minute of work time remaining. Most of the children begin to put away the manipulatives and prepare for the discussion of their solutions.

The teacher starts the discussion by asking the children to explain how they solved the first activity. Typically, there is no shortage of volunteers. When called upon, children spontaneously come to the front of the class to give their explanations, often accompanied by their partner. If a child gives only an answer, the teacher asks for an explanation of the solution process. Sometimes she asks follow-up questions to clarify the explanation or to help the child reconstruct and verbalize a solution. Occasionally, a child will become aware of a problem with his or her solution while explaining it to the class. Because of the accepting classroom atmosphere, the child does not become embarrassed or defensive but might simply say, "I messed up" and sit down. In the course of the dialogue, other children spontaneously explain why they now disagree with their own solutions or why they thought their partner was wrong but now think he or she is right. It is immediately apparent that the teacher accepts all answers and solutions in a completely nonevaluative way. If, as frequently happens, children propose two or more conflicting answers, she will frame this as a problem for the children and ask them how they think the conflict can be resolved. Children volunteer to justify particular answers and, almost invariably, the class arrives at a consensus. On the rare occasions when they fail to do so, the teacher writes the activity statement on a chalk board so that the children can think about it during the following few days. Although the discussion might continue for 15 to 20 minutes, the time is sufficient to consider only a small proportion of the activities completed by some groups – the children have much to say about their mathematics. Eventually, the teacher terminates the discussion due to time constraints. She collects the children's activity sheets and might glance through them before making them ready for distribution to parents. However, she does not grade their work or in any way indicate whether or not their answers are correct.

In the remaining ten minutes or so of the hour long lesson, the teacher introduces a whole class activity and poses one or more questions to the children. She is again nonevaluative when the children offer their solutions and, as before, attempts to orchestrate a discussion among the children.

1.3 Development Process

The teaching experiment bears certain resemblances to a type of Soviet experiment that Menchinskaya (1969) called a macroscheme: "Changes are studied in a pupil's school activity and development as he [or she] makes the transition from one age level to another, from one level of instruction to another" (p. 5). However, there is a crucial difference between our approach and that of the Soviet researchers. Typically, Soviet investigators construct the instructional materials before the experiment begins (e.g., Davydov, 1975). We, in contrast, developed samples of a wide range of possible instructional activities in the year preceding the experiment, but the specific activities used in the classroom were developed, modified, and, in some cases, abandoned while the experiment was in progress. To aid this process, two video-cameras were used to record each mathematics lesson for the school year. Initial analyses of both whole class dialogues and small group problem solving interactions focused on the quality of the children's mathematical activity and learning as they tackled specific instructional activities. These analyses, together with the classroom teacher's observations, guided the development of instructional activities and, on occasion, changes in classroom organization for subsequent lessons. Thus the processes of developing materials, conducting a formative assessment, and developing an initial explanation of classroom life were one and the same.

1.4 Research Emphases

For the sake of explication, we have separated the development and research aspects of the project. In practice, however, we found that the two frequently blended together to such an extent that the distinction became irrelevant. As we have noted, we struggled to achieve an initial understanding of classroom events as part of the development process. In doing so, we attempted to clarify situations that were problematic to us and, in the process, developed research questions. The speculative solutions and working hypotheses proposed in turn fed back to inform the development process. Thus, just like the children, we encountered problematic situations that constituted opportunities to learn as we attempted to achieve the goals of the project in the classroom.

Before describing our research emphases, it is perhaps well to state categorically that our intent has not been, is not, and cannot possibly be to prove that constructivism is right or even the way that all researchers should frame the problems of mathematics education. It is simply a way of knowing which, we believe, might open up potentially fruitful avenues of investigation. Further, we have tried to be consciously aware of the danger of forcing our observations into the conceptual boxes provided by constructivism as we understood it at the beginning of the project. As a consequence, we have had to revise and relinquish some basic assumptions that initially seemed beyond question. Thus, part of the research agenda requires that we become increasingly aware of and probe some of the weaknesses of constructivism. This, in our view, is the soundest way of contributing to the vigor of this or any paradigm.

The issues of current interest include the children's construction of increasingly abstract arithmetical objects, their emotional/moral development, the teacher's learning in the classroom, and the changing nature of the children's beliefs about the activity of doing mathematics, their role as students, and the teacher's role. There is, however, one overriding problem that touches each of these issues. Constructivism, at least as it has been applied to mathematics education, has focused almost exclusively on the processes by which individual students actively construct their own mathematical realities. Much progress has been made on this front in recent years. However, far less attention has been given to the interpersonal or social aspects of mathematics learning and teaching. Given the emphasis of the evolving subjective realities of individuals, how (to state the question in observer language) does mathematics as cultural knowledge become "interwoven" with individual children's cognitive achievements (Saxe, Guberman, & Gearhart, 1985)? In other words, how is it that the teacher and the children manage to achieve at least temporary states of intersubjectivity when they talk about mathematics? More simply still, how do children learn in such instructional situations? In our view, these are critical questions for constructivism. We will be unable to talk about the specifics of instruction in a theoretically grounded way unless we place analyses of learning within the context of classroom social interactions. Brophy and other adherents of the process-product approach are clearly trapped on one side of the chasm that currently separates research on learning from research on teaching. Constructivists are in danger of becoming trapped on the opposite side of the same divide.

As a first step in coming to grips with the problem of learning in social settings, we are currently analyzing the evolving regularities or patterns identified in both whole class and small group interactions. These patterns are, for the most part, outside the conscious awareness of both the teacher and the children and are repeatedly reconstructed in the course of interactions (Voigt, 1985). In other words, the patterns consist of coordinated sequences of actions and, at each occurrence, the development of a pattern begins anew – the enactment and the construction of a pattern are synonymous. Thus, although the teacher and children do not have a "blueprint" of the interaction patterns, each knows how to act appropriately in particular situations. The patterns reveal the largely implicit social norms negotiated by the teacher and children, the norms that constitute the social reality within which they teach and learn mathematics. Following Voigt (1985), we are analyzing the patterns and associated norms in terms of both the implicit, taken-for-granted obligations that the teacher and children accept in particular situations and the expectations they have for each other. The investigation of how they mutually construct the social reality of the classroom will therefore elucidate their beliefs about their own and each others' roles and document how they created a "problem-solving atmosphere" (Silver, 1985).

At the same time, we are interested in how the evolving network of obligations and expectations influences individual children's construction of mathematical knowledge. The relevance of this question is apparent when one observes with Balacheff (1986) that "most of the time, the pupil does not act as a *theoretican* but as a *practical man*. His job is to give a solution to the problem the teacher has given him, a solution that will be acceptable with respect to the classroom situation" (p. 12). Thus, the social norms constrain what is problematic for children and what might count as an acceptable solution. However, it is while the children attempt to complete the mathematical activities in groups and discuss their solutions in the whole class setting that the social norms are renegotiated. This leads us to the contention that neither the cognitions of individuals nor the mutual constructed network of obligations and expectations are primary; we find it impossible to give an adequate explanation of one without considering the other.

On the other hand, the social norms and consensually sanctioned mathematical knowledge are created, regenerated, and modified by the coordinated actions of the teacher and children as individuals. The norms and consensually sanctioned knowledge do not exist independently of these actions (except in the mind of the analyst). On the other hand, each individual's understanding of the norms and knowledge contrains his or her activity in the classroom. The norms (and consensually constructed mathematics knowledge) constrain the activity that creates the norms. Conversely, individuals' activity creates the norms that constrain that activity.

In this formulation, the individual and the social are interdependent, in that one does not exist without the other. We thus acknowledge that social context is an integral aspect of an individual's cognitions without reifying mathematics as a ready-made body of cultural knowledge that is somehow internalized from without by individuals. Neither (the observer's) physical, mathematical, or social realities are taken as solid, immutable bedrock upon which to anchor an analysis of learning and teaching.

II. EXAMPLES FROM THE CLASSROOM

We present our current, admittedly fragmentary understanding of classroom life in the following sections. At the same time, the protocols and narratives of specific episodes that support the analysis exemplify the kinds of interactions that typified the activities of learning and teaching mathematics in the classroom.

2.1 Whole Class Interactions

The teacher's overall intention as she led whole class discussions was to encourage the children to verbalize their solution attempts. Such dialogues give rise to learning opportunities for the children as they attempt to reconstruct their solutions, understand alternative points of view, resolve conflicts between incompatible solution methods, and so forth. The teacher also had learning opportunities in that she had to decenter and "see" things from the children's perspectives when she helped them to say what they wanted to say. However, the teacher's expectation that the children should verbalize how they had actually interpreted and attempted to solve the instructional activities did not fit with the expectations the children had developed on the basis of their prior experiences of class discussions in school. During both first grade and second grade with the exception of mathematics, such dialogues were typically initiated with the intention of steering or funnelling (Voigt, 1985) the children towards an officially sanctioned intepretation or solution (Wood, Cobb, & Yackel, 1988). As a consequence, class discussions in subject matter areas other than mathematics were situations in which the children felt obliged to try and infer what the teacher had in mind rather than to articulate their own understandings. The project teacher therefore had to exert her authority in order to help the children reconceptualize both their own and her role during mathematics instruction. In effect, she had to actively teach the children that she had different expectations for them when they did mathematics. To this end, she *initiated* the mutual construction of expectations and obligations in the classroom. In doing so, she simultaneously had to accept certain obligations for her own actions. If she expected the children to honestly express their current understandings of mathematics, then she was obliged to accept their explanations rather than to evaluate them with respect to an officially sanctioned solution method. Thus, the teacher had obligations to the children, just as they did to her. This evolving, interlocking network of obligations and expectations was beyond the conscious awareness of both the teacher and the children, but nevertheless was mutually constituted. The teacher and the children initially negotiated obligations and expectations at the beginning of the school year, which subsequently made possible the smooth functioning of the classroom for the remainder of the school year. Once established, this mutually constructed network of obligations constrained classroom social interactions in the course of which the children constructed mathematical meanings as they attempted to achieve their goals (Blumer, 1969). Thus the patterns of discourse served, not to transmit knowledge (Mehan, 1979; Voigt, 1985), but to provide opportunities for children to articulate and reflect on their mathematical activities.

The teacher's and students' mutual construction of social as well as mathematical realities is reflected in the dual structure of classroom dialogues. In one conversation, they talked about mathematics whereas in the other they talked about talking about mathematics. As these two conversations were conducted at distinct logical levels (Bateson, 1973), one in effect setting the framework for the other, seemingly contradictory statements such as "the teacher exerted her authority to enable the children to express their viewpoints" make sense. As in a traditional classroom, the teacher was very much an authority in the classroom who attempts to realize an agenda. The difference resides in the way that she translated her authority into action (Bishop, 1985).

In the following episode that occurred during the first day of school, the discussion centers on word problems that are shown on an overhead projector.

T: Take a look at this problem. The clown is first in line. Which animal is fourth? Peter.
Peter: The tiger.
T: How did you decide the tiger? .... Would you show us how you got the fourth?
Peter: (Goes to the screen at the front of the room.) I saw the clown and then ... (He counts the animals.) Oh, the dog [is fourth]. (Hesitates) Well, I couldn't see from my seat. (He looks down at the floor.)
T: OK. What did you come up with?
Peter: I didn't see it. (He goes back to his seat quickly.)

The teacher realized that in making Peter obliged to explain his solution, she put him in the position of having to admit that his answer was wrong in front of the entire class. Peter's response to the situation was to offer an excuse for his error, "I couldn't see it ..." Peter's concern with social comparison became manifest by his construal of the situation as warranting embarrassment (Armon-Jones, 1986). This construal confounded the teacher's intention that the children should feel free to publicly express their own solutions to problems. For her purposes, it was vital that children feel no shame or embarrassment when they present erroneous solutions in front of others. Crucially, the teacher immediately initiated a second conversation by talking about talking about mathematics as she suggested an alternative construal of the situation.

T: That's okay Peter. It's all right. Boys and girls even if your answer is not correct, I am *most* interested in having you think. That's the important part. We are not always going to get answers right, but we want to try.

The teacher was directive in her comments. She expressed her expectations by telling the children how she as an authority interpreted the situation. She emphasized that Peter's attempts to solve the problem were appropriate in every way, and simultaneously expressed to the other children her belief that what is important in this class is thinking about mathematics, not just getting right answers. She then terminated this conversation and returned to the first.

T: All right, did anybody else come up with a different way to do the problem?

In this episode, the project teacher practiced what Schon (1985) called "knowledge-in-action." She did not follow a set of prescribed rules as she conducted this and other dialogues. Instead, as a practitioner, she made sense of a problematic situation by reframing the situation.

The example makes clear that constructivist teaching does not mean that "anything goes" or that the teacher gives up her authority and abrogates her wider societal obligations. Because mathematical meaning is inherently dependent on the construction of consensual domains, the activities of teaching and learning must necessarily be guided by obligations that are created and regenerated through social interaction. Critically, the teacher attempted to achieve her agenda by initiating and actively guiding the mutual construction of obligations and expectations in the classroom. In doing so, she attempted to invite actions that she considered appropriate while hindering others. These invitations and her overall agenda reflected her beliefs that mathematical learning is a problem-solving activity and that children's understandings should be respected. As a consequence of the teacher's interventions, the realized dialogue patterns contrast sharply with those of a typical classroom (Mehan, 1979; Stake & Easley, 1978; Voigt, 1985). The teacher exercised her authority in different ways in each of the two conversations that intertwined in the dialogues. When she and the children talked about talking about mathematics, the teacher typically initiated and attempted to control the conversation. However, when they talked about mathematics, she limited her role to that of orchestrating the conversation. The following episode taken from a dialogue that occurred later in the year again illustrates her authoritarian role with respect to the classroom social norms.

T: Now another thing I noticed was happening and it is something I *don't* like and I *don't* want to hear ... and it's these two words. (She writes "That's Easy on the board and draws a circle around the phrase.) These words are no-no's starting today. What are these two words Mark?

Mark: That's easy.

T: ... I've had kids come up to me and say, "Oh, that's easy!" Maybe I look at it and say, "I don't think that's very easy." How do you think that's going to make me feel?

It is evident as she listened to several children's suggestions that she was looking for a specific answer. Finally she stated, "You are not saying the word I am looking for." She continued to fish for a specific interpretation of the situation without success, and ultimately directly told the children.

T: ... it hurts my feelings when someone says. "Oh that's easy!" (She points to the words on the board.) When I am struggling and trying so hard, it makes me feel kind of dumb or stupid.

I Because I am thinking, gosh, if it's so easy, why am I having so much trouble with it?

Once she had explicitly told the children the officially sanctioned interpretation of the situation, she terminated the social conversation and turned to talking about mathematics.

T: That's easy is a real put down. It's like if I think it's easy, then you must not be very smart, because it is not easy for you ... OK. The first activity is balances.

The teacher's highly directive intervention is consistent with her agenda for the classroom. By outlawing actions that might make others feel stupid, the teacher nurtured the sense of trust that was essential if the children were to talk publicly about their mathematics. She demonstrated that if the children accepted the obligation of expressing their mathematical understandings then she, as an authority, was obliged to protect them. Her efforts bore fruit in that even the most conceptually immature child seemed to feel that his thinking was valued and respected, and as a consequence continued to participate actively in whole class discussion throughout the year.

When a conflict existed between the two conversations, the teacher always gave priority to the social (Stake and Easley, 1978). She was right to do so in that the norms established in the course of the social dialogue made possible the conversations in which children felt free to take the initiative as they talked about mathematics. The teacher was then able to adopt a nonauthoritarian, nonevaluative role when the children explained their solutions. For the most part, she orchestrated their contributions to the class discussions. The manner in which the children felt free to take the initiative is illustrated in the following episode which occurred near the end of the school year. The instructional activities for the previous two days provided a setting in which to discuss the meanings of fraction symbols such as 1/4, 1/2, 1/6, etc. The discussion prior to this episode was about the meaning of 7/8ths. The episode begins when John, who had previously stated that he has a problem he has been thinking about, came to the front of the room where the teacher, the overhead projector, and the screen were located.

John: But um ... what about if it was one and one? What would that be?
T: Like this? (She writes 1/1). Good question. Or what if you had this over this? (She writes 4/4 and 6/6.) What does that mean?
John: (Starts talking to the teacher while tracing two circles on the screen with his hand.) Put a circle down.
T: O.K. Put a circle down (she draws a circle on the overhead, then looks at John).
John: Um like ... what I'm thinking is one and one. (He points to 1/1.) How could you make like ...? (He looks at the teacher.)

The teacher's response at this point deviated distinctly from typical patterns of classroom interaction in that she did not offer John an official explanation but instead she continued the turn-taking dialogue by asking him a question that placed him under the obligation of thinking through a solution.

T: How could you show one and one, one over one?
John: (Looks at the teacher and says nothing.)
Ann: (Interjects) Just fill it all in.
Mark: Yeah! That's one piece.
T: OK. If we remember what a fraction tells us, John ...
John: (Interrupts the teacher excitedly.) But it's the whole thing (he gestures with his hands, indicating a larger item), not just one piece (holds up one finger).
T: So, how many can we fill in?
John: The whole thing!

The conversations in which children discussed mathematics on *their* terms gave rise to problematic situations for them which constituted opportunities for reflection and the construction of mathematical knowledge. These interactions were characterized by a genuine commitment to communicate. All the participants came to assume that any contribution to the dialogue made sense to the speaker (Rommetveit, 1985). Participation in the discussions therefore involved a genuine attempt to exchange points of view and encouraged the children to distance themselves from on-going activity (Sigel, 1981).

The manner in which the patterns of interactions evolved during the school year exemplifies the development of what Silver (1985) called a "problem-solving atmosphere." The teacher's priority was to establish a nonevaluative, cooperative setting in which children could publicly verbalize and reflect on their mathematical ideas without risk of social comparison or public embarrassment.

The obligations and expectations mutually constructed during whole class discussions also provided a framework for the children's activity as they worked in small groups, in that they were expected to solve problems in a cooperative manner and to respect each others' efforts. However, the children still had to negotiate obligations and expectations in the small group setting that would constitute a productive working relationship.

2.2 Small Group Interactions

The children attempted to complete instructional activities in groups of two, or occasionally three, prior to the whole class discussions. From the beginning of the year the teacher used her authority to emphasize two major responsibilities for each small group, namely that they cooperate to complete the activities and that they reach a consensus. Ideally, consensus is attained when the children develop a mutually acceptable method of solution. In a more limited sense, consensus is reached when the children agree on a common answer, albeit via different methods of solution. Thus, the children had two distinct types of problem to solve. The first concerned the mathematical problems that arose as they attempted to complete the instructional activities, whereas the second was that of negotiating a viable relationship that would make it possible jointly to solve their mathematical problems.

As with the whole class setting, the teacher was very explicit about what she expected of the children as they worked to solve both types of problem. For example, she attempted to place the children under the obligation of persisting and thinking their problems through for themselves rather than trying to complete as many activities as possible. Thus, she assured the children "it's OK to complete one problem or five problems. If you don't get one finished, don't worry." The children not only demonstrated by both word and deed that they understood and accepted the teacher's expectations (gaining personal satisfaction by solving difficult problems, persistence, and task-involvement came to characterize small group work), but felt free to extend and further emphasize them. This is illustrated by the following dialogue which occurred in the whole class setting at the beginning of a mathematics lesson. This episode occurred on the first day following two weeks of winter holiday.

Lois: How many problems are there?
Teacher: There are eight. So the most you can do is eight. Some of them are very hard though, so you may only get two done. Is that OK if you only get two done?
Students: (in unison): Yeah.
Teacher: You bet it is ... if you only get one done because it was so hard and you worked so hard on it that it was the only one you got, that's OK too.
Adam: If you get none done, but you're still working on it, it's OK too.

By contrasting her expectations with the typical classroom obligation of completing a prescribed number of tasks, the teacher attempted to place the children under the obligation of persisting to resolve what they found problematic. Adam's elaboration provides clear evidence that children understood the teacher's expectations and accepted the consequences. Even the most extreme case, completing none, was acceptable provided the children met the obligation of striving to solve their problems.

The above discussion illustrates that the teacher was very much an authority when she told the children what she expected of them as they tried to solve their mathematical problems. She was even more explicit about what she expected socially, as is illustrated in the following continuation of the episode. Here, the teacher reminded the children of their social obligations as they worked in small groups.

Teacher: Any questions before we get started? OK. You're going to cooperate. You're going to work with your partner. And if *you* figure out the answer and your partner is looking at you like, "How in the world did you get that answer?" it's going to be up to you and your partner to work it out and understand it together. *Then* you can get another [problem] card, but not until then.

In this directive the teacher attempted to place the children under a dual obligation. They were expected to work cooperatively and, at a minimum, to develop solutions that produced the same result. Ideally, they should also develop mutually acceptable methods of solution. The two obligations are at very different levels of complexity. At the basic level, children must negotiate a way to work with each other and must share materials including the activity pages (each small group received only one copy.) Cooperation at this level includes listening to and attempting to understand each other's comments and, more generally facilitating rather than hindering each other's mathematical activity. The second level of cooperation, working together to construct a mutually acceptable solution, is much more complex. It requires that students communicate about mathematics, verbally or otherwise.

The social problem of cooperating at the basic level always has priority in the sense that it makes the construction of mutually acceptable solutions possible. The teacher played an important role in facilitating cooperation at this level (Wood & Yackel, 1988). As the teacher observed small group activity, she was able to identify those groups which were having difficulty at this basic level. Comments such as, "if you put the paper here in the middle then both of you can see it" and "Listen to what your partner is saying" exemplify the interventions the teacher made to encourage cooperation. If children were unable to resolve the problem of achieving basic cooperation over several weeks, even with the teacher's intervention, they were assigned new partners. In the few cases where basic cooperation was not achieved, there were underlying reasons for failure such as extremely immature social communication skills, high ego-involvement, or large discrepancies in mathematical conceptual level. As the year progressed, the children's social development alleviated the difficulties of establishing cooperative relationships in the small group setting. By the end of the year, instances of failure to cooperate at the basic level were isolated and were typically resolved by the children themselves within the group.

The problem of cooperating at the level of developing a mutually acceptable solution is much more demanding. Nevertheless such cooperation became the norm in the classroom. At this level of cooperation, the children have to appreciate what each finds problematic and coordinate their mathematical problem solving activities. Such interactions naturally give rise to opportunities for reflection and cognitive reorganization. As an illustration, extracts of a dialogue between two children as they solved a sequence of multiplication tasks are presented. The children had just successfully completed the task $5 \times 4 =$ ____. (The children's use of the term "sets" in talking about multiplication is a result of the teacher's use of the term when she first introduced "×" as the mathematical symbol for multiplication.)

$10 \times 4 =$ ___.

John: It's five more sets [of 4]. Look. Five more sets than 20.
Andy: Oh! 20 plus 20 is 40. So its gotta be 40. No?
John: Yeah!
Andy: No. 4, 8, 12, 16, 20,24, 28 (keeps track on his fingers).
John: 40.
Andy: 40.
John: Yeah, I know ... cause ten fours make 40.
Andy: Like five fours make 20.
John: Four sets of ten makes 40. Just turn it around.
Andy: Five sets of fours make 20 and so five more than that.
John: Yeah, just turn it around. Just turn it around.
Andy: 5 times 4 is 20, so 20 more than that makes 40.
John: Just switch them around.

Notice that Andy constructed his solution by elaborating John's initial comment that "It's 5 more sets [of four]." However, in his final explanation John developed an alternative solution method. Andy's initial statement that 40 is the answer was followed by John's exclamation "Yeah!" John's subsequent verbalizations suggest that once he heard "20 plus 20 is 40" he reconceptualized the problem as four tens, rather than as 20 and 5 more fours, and then successfully justified that conceptualization. This inference is supported by the frequency and enthusiasm with which he repeated "Just turn it (or switch them) around." Thus, even though each child eventually conceptualized the problem differently their initial verbalizations were instrumental in the development of the other's solution. This illustrates an extremely productive level of cooperation in which each partner facilitated the conceptual activity of the other.

$9 \times 4 =$ ____

John:	Just take away 4 from that [10 × 4].
Andy:	36
John:	(pause) Yeah!

In this brief dialogue we see a different form of cooperation. One partner, John, suggested a solution process and the other, Andy, carried it out. The pause before John's final agreement indicates that he verified Andy's answer. In the following dialogue the children agreed on an answer but constructed different methods of solution.

$8 \times 4 =$ _____

John:	Look! Look! Just take away 4 from that [9 × 4] to get that [8 × 4]. See! Just take away 4 from there [9 × 4].
Andy:	Just take 8 away from that [10 × 4].
John:	No. Take away 4 from there [9 × 4].
Andy:	Take 8 away from that [10 × 4]. That makes 32.
Observer:	Did you do it the same way as Andy?
John:	Yeah, but I used that one [9 × 4]. Take away 4. It makes 32.

John's final comment makes it evident that he was consciously aware of the similarities and differences between his and Andy's methods. Although they constructed different methods, each was satisfied that his was viable. This was sufficient to meet the obligation of reaching consensus. As the dialogue continued, the children encountered a situation where they disagreed on the answer.

$8 \times 5 =$ ___

Andy:	Five more than that [8 × 4] is 37.
John:	Eight sets of 4. Eight sets of 5.
Andy:	No. 9, 39, I think.
(Both children pause to reflect for a few moments.)	
John:	(very excitedly) It's 40.
Andy:	It is?
John:	Yeah, its 40! Yeah, look!
Andy:	No, it's 39.
John:	5, 10, 15, 20, 25, 30, 35, 40, 45, oops. Take away that last one. It's 40.
Andy:	5, 10, 15, 20, 25, 30, 35, 40, 45.
John:	Huh?
Andy:	Wait. 40. Yeah, 40.

On this task, Andy first attempted to modify his previously successful method of solution and found eight fours plus five. Apparently John's remark, "Eight sets of 4. Eight sets of 5," led Andy to reflect on this activity. His next answer,

39, indicates that he modified his initial conceptualization. Andy then challenged John's answer of 40, and John was obliged to justify his solution. Andy immediately adopted John's method but his conceptualization of it is open to question as he simply repeated exactly what John had said, including the error of going up to 45. But John's "Huh?" prompted further reflection and they finally agreed on the answer. In this instance the problem of achieving a consensus provided the basis for their mathematical activity.

The final task in this excerpt is 7 × 5 = ____. The dialogue is very brief.

Andy: Seven sets of 5.
John: Oh, it's just 5 lower than that. It's ...
Andy and John: (simultaneously) 35.
John: We both said it at the same time.

John appeared amused and pleased that they said the answer simultaneously, as though that were the height of cooperation.

The above excerpts illustrate that the coordination of mathematical activity can take a variety of forms. Notice that nowhere in the dialogue did either partner make explicit reference to their obligation to cooperate. (Typically such comments are made only when cooperation breaks down.) Yet it is apparent that in every instance they persisted until each had constructed a viable method that produced a common answer. The smooth flow of their discussions indicates that they had mutually constructed a viable network of obligations and expectations. For example, each knew when it was appropriate to explain or to justify a solution without being explicitly asked to do so. The two children's mathematical discussions, which are characteristic of this level of cooperation, gave rise to a variety of learning opportunities that are unique to the small group instructional strategy.

The level of cooperation illustrated by John and Andy was made possible by their ability to make sense of each other's mathematical activity. Cases where a child, for one reason or another, is unable to make sense of his or her partner's mathematical activity make it impossible for such cooperation to occur. This in turn makes it more difficult for the children to cooperate at the basic social level. In the episode presented below the method used by one partner was incomprehensible to the other. Consequently, cooperation at the level of developing a mutually acceptable problem solution was impossible. Each worked independently to complete the instructional activities because they were unable to talk to each other about their mathematics. What dialogue there was, necessarily focused on whether or not they had the same answer and on the more basic level of cooperation. In this episode the partners were working on a multiplication activity. One child was using an (incorrect) algorithm to solve the problems. Her partner, unable to make sense of her approach, chose to solve the problems by drawing circles of tally marks and then counting. As a consequence, it took him longer to generate an answer to each task.

5 × ___ = 35.

Adam: Will you wait for me? Five times something equals 35. I already know the answer to that. Five times five, I think. Let me check it again. (At this point Adam draws five circles of five.) Seven.

$4 \times 7 =$ ___

Adam: Four times seven equals something. Let me draw these here. You can go ahead. (He draws four groups of seven.) Oh, one more and then I'll be caught up.
Ann: (Writes 22 as her answer on the activity page.) Do you agree?
Adam: Hold on. You're going ahead.
Ann: I hate waiting for you.
Adam: I got 28 on that one but hold on.
Adam checks the answer 28 again and they agreed to give that as the answer.

Notice in the above dialogue that the children did not discuss their mathematics. At best, they could compare answers and share the activity page and any other materials they had. As a consequence, they continually had to resolve the social problem of synchronizing their independent activities. Adam, for example, continued to insist that Ann should wait for him and that they should at least work on the same activity. Ann reluctantly obliged; she too understood that they were expected to work cooperatively and to agree on answers. Despite their different mathematical understandings, they attempted to maintain a basic level of social cooperation. To do so, they had to continually negotiate cooperation at this level verbally. This dialogue contrasts sharply with that between John and Andy, who cooperated fully but made no verbal reference to it.

In the project classroom the social relationships the children negotiated were influenced by, but not determined by, their level of conceptual understanding of mathematics. For example, partners who were at approximately the same conceptual levels could readily make sense of each other's mathematical thinking. Thus, cooperation at the level of developing mutually acceptable problem solutions was facilitated. John and Andy exemplify this type of relationship. Partners at widely different conceptual levels could cooperate only if the more advanced partner operated far below his or her capabilities and discussed mathematics at the level of the less advanced partner, or if partners agreed to work independently on the instructional activities and compare answers. Although the first option presents opportunities for the more advanced child to serve as a tutor and can give rise to learning opportunities, in general it denies that child opportunities for conceptual advancement. In the project classroom partners who evidenced such a relationship were reassigned.

Some children who were at very different conceptual levels did develop very productive working relationships. For example, in one pair the child who was less advanced conceptually was very task-involved and continually requested assistance from his partner. It was common for these children to work individually on the mathematics tasks and then compare their answers. Dialogue in this group, as in the Adam/Ann episode presented earlier, frequently focused on negotiating a basic level of cooperation.

The previous discussion and examples illustrate that the social interactions between partners influence their mathematical activity and give rise to learning opportunities. The interactions are in turn influenced by their conceptual understanding and by the network of obligations and expectations mutually constructed both within the group and in the wider whole class setting. Our observations in the project classroom indicate that, as early as second grade, children are remarkably successful at working together to complete instructional activities. In doing so, they solve not only their mathematics problems but also the problem of how to work together.

III. CONCLUSIONS

The teaching experiment was reasonably successful in that we were able to satisfy the school district administrators' expectations while allowing the children to solve their mathematical problems in ways that were acceptable to them. It also became apparent that the children's abilities to establish productive social relationships and to verbalize their own thinking improved dramatically as the year progressed. However, most observers seemed to attribute the greatest significance to the emotional tone of the classroom. In general, the children were enthusiastic, persistent, did not become frustrated, and experienced joy when they solved a personally challenging problem. We have, for example, been unable to identify a single instance when a child evidenced frustration while problem solving during the second semester of the school year. Further, the children's persistence was such that we found it necessary to allocate three consecutive one hour class periods to general problem solving whenever these instructional activities were used. The children requested additional time because they wanted to continue working on an activity they had been unable to complete during an entire class period.

We have tried to make the point that the teacher was very much an authority in the project classroom, albeit a benevolent one. Her success in placing the children under certain obligations while readily accepting compatible obligations for herself was crucial to her effectiveness. As a constructivist teacher, she did not merely refrain from carrying out certain activities characteristic of traditional teachers and relinquish her authority. Rather, she expressed that authority in action by initiating the mutual construction of certain obligations and expectations. In doing so, she influenced the children's beliefs about both the nature of the activity of doing mathematics and their own and the teacher's roles in the classroom. Above all else, the obligations and expectations that were established constituted a trusting relationship. The teacher trusted the children to resolve their problems and they trusted her to respect their efforts. This trust is, in our opinion, the most important feature of constructivist teaching.

REFERENCES

Armon-Jones, C. (1986). "The Thesis of Constructionism". In R. Harré (ed.), *The Social Construction of Emotions* (pp. 33–56). Oxford: Blackwell.

Balacheff, N. (1986). "Cognitive Versus Situational Analysis of Problem- Solving Behavior." *For the Learning of Mathematics*, 6 (3), 10–12.

Bateson, G. (1973). *Steps To An Ecology Of Mind*. London: Paladin.

Bishop, A. (1985). "The Social Construction of Meaning - A Significant Develoment for Mathematics Education?" *For The Learning of Mathematics*, 5, (1), 24–28.

Blumer, H. (1969). *Symbolic Interactionism*. Englewood Cliffs, NJ: Prentice-Hall.

Brophy, J. (1986). "Teaching and Learning Mathematics: Where Research Should Be Going." *Journal for Research in Mathematics Education*, 17, 323–346.

Carpenter, T.P. (1983). "Epistemology and Research in Mathematics Education: A Reaction". In J.C. Bergeron & N. Herscovics (eds.), *Proceedings of the Fifth Annual Meeting of the North American Chapter of the International Group for the Psychology of Mathematics Education*. Montreal: PME-NA.

Cobb, P. (1986). "Concrete Can Be Abstract: A Case Study." *Educational Studies in Mathematics*, 17, 37–48.

Confrey, J. (1987, July). *The Current State of Constructivist Thought in Mathematics Education*. Paper presented at the annual meeting of the International Group for the Psychology of Mathematics Education, Montreal.

Davydov, V.V. (1975). "The Psychological Characteristics of the "Prenumerical" Period of Mathematics Instruction." In L.P. Steffe (ed.), *Soviet Studies in the Psychology of Learning and Teaching Mathematics* (Vol. 7, pp. 109–205). Stanford, CA: School Mathematics Study Group.

Erickson, F. (1986). "Qualitative Methods in Research On Teaching". In M.C. Wittrock (ed.), *The Handbook of Research on Teaching*, 3rd edition. NY: MacMillan.

Mehan, H. (1979). *Learning Lessons: Social Organization in the Classroom*. Cambridge, MA: Harvard University Press.

Menchinskaya, N.A. (1969). "Fifty Years of Soviet Instructional Psychology". In J. Kilpatrick & I. Wirszup (eds.), *Soviet Studies in the Psychology of Learning and Teaching Mathematics* (Vol. 1, pp. 5–27). Stanford, CA: School Mathematics Study Group.

Rommetveit, R. (1985). "Language Acquisition as Increasing Linguistic Structuring of Experience and Symbolic Behavior Control." In J.V. Wertsch (ed.), *Culture, Communication, and Cognition* (pp. 183–205). Cambridge: Cambridge University Press.

Rumelhart, D.E. and D.A. Norman (1981). "Analogical Processes in Learning". In J.R. Anderson (ed.), *Cognitive Skills and Their Acquistion* (pp. 335–359). Hillsdale, NJ: Erlbaum.

Saxe, G.B., S.R. Guberman and M. Gearhart (1985, April). *The Social Context of Early Number Learning*. Paper presented at the annual meeting of the American Educational Research Association, Chicago.

Schon, D. (1985). *The Design Studio: An Exploration of Its Traditions and Potentials*. London: RIBA Publications.

Sigel, I.E. (1981). "Social Experience in the Development of Representational Thought: Distancing Theory." In I.E. Sigel, D.M. Brodzinsky, and R.M. Golinkoff (eds.), *New Directions in Piagetian Theory and Practice* (pp. 203–217). Hillsdale, NJ: Lawrence Erlbaum Associates.

Silver, E.A. (1985). "Research on Teaching Mathematical Problem Solving: Some Underrepresented Themes and Needed Directions." In E.A. Silver (ed.), *Teaching and Learning Mathematical Problem Solving: Multiple Research Perspectives* (pp. 247–266). Hillsdale, NJ: Lawrence Erlbaum.

Stake, R.E. & J. Easley (1978). *Case Studies in Science Education* (Vol. 2). Urbana, IL: University of Illinois Center for Instructional Research and Curriculum Evaluation.

Steffe, L.P., P. Cobb and E. von Glasersfeld (1988). *Young Children's Construction of Arithmetical Meanings and Strategies*. NY: Springer-Verlag.

Steffe, L.P., E. von Glasersfeld, J. Richards, & P. Cobb (1983). *Children's Counting Types: Philosophy, Theory, and Application*. NY: Praeger Scientific.

Thompson, P. (1985). "Experience, Problem Solving, and Learning Mathematics: Considerations in Developing Mathematics Curricula." In E.A. Silver (ed.), *Teaching and Learning Mathematical Problem Solving: Multiple Research Perspectives* (pp. 189–236). Hillsdale, NJ: Erlbaum.

Voigt, J. (1985). "Patterns and Routines in Classroom Interaction". *Recherches en Didactique des Mathematiques*, 6, 69–118.

von Glasersfeld, E. (1983), "Learning as a Constructive Activity." In J.C. Bergeron & N. Herscovics (eds.), *Proceedings of the Fifth Annual Meeting of the North American Chapter of the International Group for the Psychology of Mathematics Education* (Vol. 1, pp. 41–69). Montreal: PME-NA.

Wood, T., P. Cobb, & E. Yackel (1988, April). *The Influence of Change in Teacher's Beliefs About Mathematics Instruction on Reading Instruction*. Annual meeting of the American Educational Research Association, New Orleans.

Wood, T. & E. Yackel (1988, July). *Teacher's Role in the Development of Collaborative Dialogue Within Small Group Interactions*. Paper presented to the Sixth International Congress on Mathematical Education, Budapest.

THE CONSTRUCTIVIST TEACHING EXPERIMENT: ILLUSTRATIONS AND IMPLICATIONS[1]

LESLIE P. STEFFE

University of Georgia, Athens, GA

In an epistemology where mathematics teaching is viewed as goal-directed interactive communication in a consensual domain of experience, mathematics learning is viewed as reflective abstraction in the context of scheme theory. In this view, mathematical knowledge is understood as coordinated schemes of action and operation. Consequently, research methodology has to be designed as a flexible, investigative tool.

The constructivist teaching experiment is a technique that was designed to investigate children's mathematical knowledge and how it might be learned in the context of mathematics teaching (Cobb & Steffe, 1983; Hunting, 1983; Steffe, 1984). In a teaching experiment, the role of the researcher changes from an observer who intends to establish objective scientific facts to an actor who intends to construct models that are relative to his or her own actions.

I. ROLES OF THE RESEARCHER IN A TEACHING EXPERIMENT

A distinguishing characteristic of the technique is that **the researcher acts as teacher**. Being a participant in interactive communication with a child is necessary because there is no intention to investigate teaching a predetermined or accepted way of operating. The current interest always lies in hypothesizing what the child might learn and finding ways and means of fostering this learning. Based on current interpretation of the child's language and actions, the experimenter makes decisions concerning situations to create, critical questions to ask, and the types of learning to encourage. These on-the-spot decisions represent a major modus operandi in teaching experiments and the researcher has the responsibility for making them.

Beyond acting as teacher, another role of the researcher is to analyze the knowledge involved in teaching. The researcher must build what Hawkins (1973) called a map and what I call a model of each child's mathematical knowledge. Toward this end, the teaching experiment is primarily an exploratory tool, derived from Piaget's clinical interview and aimed at investigating what might go on in children's heads. This process involves formulating and testing hypotheses about various aspects of the child's goal-directed mathematical activity in order to learn what the child's mathematical

E. von Glasersfeld (ed.), Radical Constructivism in Mathematics Education, 177–194.

knowledge might be like. It is my belief that the researcher can best formulate and test hypotheses and interpret the results of the tests in intense interactive communication with the child, so that a close personal and trusting relationship can be formed.

The formulation and tests of hypotheses involve initiating probes that might stretch the child to the limits of his or her conceptual adaptability and endurance. So, the researcher must strive to develop an operating context in which the social meanings of the involved language and actions are negotiated by the participants (Cobb & Steffe 1983, pp. 84ff). Above all, the activity of doing mathematics should be viewed with a playful attitude and confidence. To promote a playful attitude, the researcher might, for example, develop the expectation in a child that he or she can create situations for the teacher to solve. Such situations can not only reveal the child's current level of development but they also tend to encourage the child to develop the confidence occasionally to control the interactive communication and to make decisions concerning the situations he or she wishes to explore.

Finally, because the teaching experiment involves experimentation with the ways and means of influencing children's knowledge, it is more than a clinical interview. It is directed toward understanding the progress children make over extended periods of time, and one of the main goals is to formulate a model of learning the particular content involved. Consequently, each teaching episode is video taped and the recorded material is retrospectively reviewed after the teaching experiment. Along with the experiences of the researcher that led up to this review, models of the constructive activity of the children over the duration of the teaching experiment are formulated (Steffe, Cobb, & von Glasersfeld, 1988). It is crucial to understand the modeling process as an organizing activity analogus to what Treffers calls mathematizing (Treffers, 1987, p. 51). The basic and unrelenting goal of a teaching experiment is for the researcher to learn the mathematical knowledge of the involved children and how they construct it. In the following sections, I intend to illustrate what I mean by learning the mathematical knowledge of children and to make explicit some of its implications for mathematics education.

II. MATHEMATICS OF CHILDREN

One of the fundamental results of Piaget's genetic epistemology (Piaget, 1970) is that the roots of mathematical knowledge can be found in the general coordination of the child's actions. From a constructivist point of view, "the essential way of knowing the real world is not directly through our senses, but first and foremost through our material or mental actions" (Sinclair, 1990). There are individual actions like throwing, pushing, lifting, and the like. Mathematical knowledge is based on coordinations of such actions into organized action patterns to achieve some goal. I take these goal-directed action patterns to be what Piaget meant by sensory-motor schemes (Piaget, 1980b).[2] The mathematical knowledge of children can be understood in terms of goal-directed action patterns if "action" is taken to refer to mental as well as to physical action. These mental actions constitute operations – interiorized action patterns – and the involved schemes are operative rather than sensory-motor. From the perspective of how school mathematics is viewed today, this perspective on children's mathematical knowledge is revolutionary, and, if accepted, would transform mathematics education in a fundamental way.

To illustrate what I mean by "operative scheme", let us take how an eight-year-old child, Maya, understood division and multiplication. Maya was a participant in my most recent teaching experiment (Steffe & von Glasersfeld, 1985). When it is the goal to specify and to chart modifications of schemes, it is often advantageous to teach individual children. Choosing to work in these laboratory conditions should not be construed to mean that I view constructivism as being restricted in its implications to teaching individual children. Even in that case, interactive mathematical communication is the foundation of the teaching experiment. This is illustrated in the following protocols which document how I interpreted Maya's language and actions as well as the subsequent decisions made. The results of implementing the decisions are also documented and a qualitative analysis of what Maya learned is provided. The protocols demonstrate how essential social interaction is in cognitive construction (Piaget, 1964; von Glasersfeld, 1990).

2.1 *Maya's Schemes for Dividing and Multiplying*

In one of the teaching episodes with Maya, I presented her with 21 numeral cards in a row and then hid them from her view. "T" is used in the protocol to indicate my language and actions and "M" is used to indicate Maya's .

T: If you start from there (the beginning of the covered row) and take three cards at a time to make a pile, I wonder how many piles of three you could make? M: (Sits silently in deep concentration for approximately two minutes) Seven! W(Witness): When you counted, what did you say? M: 21, 20, 19 – that would be one; 18, 17, 16 – that would be two; etc.

Maya's method for dividing by three was to count backward by one, take each trio of number words as a unit, and count those units of three as she made them. This method was not suggested to her by me and differed substantially from the computational algorithm her classroom teacher had tried to teach her. It was an operative scheme she independently used to solve what adults commonly call "quotative division problems".

It is too strong to claim that Maya's material counting actions constituted her concept of division because of the way she organized counting by one into units of three. She intentionally counted to reach a goal. So, I looked further for the operations Maya might have used in assimilation[2] of the task. The available evidence was that Maya seemed to use her counting scheme and her concept of "three". My hypothesis was that her concept of division was quite different from what adults call "quotative division".

As a concept, quotative division would have involved an *awareness* of the result of making, say, units of three using the 21 individual units, a unit containing units of three, and a method for finding the numerosity of the containing unit before counting. Because Maya seemed to be aware of the units of three that she made *as she counted backward*, my question was whether she was aware of the containing unit before she started counting. To investigate this question, I asked Maya to formulate the results of her dividing scheme into a multiplication problem. If she was aware of the structure of her results of counting, that would be an indication she had a concept of quotative division.

T: Can you give me a multiplication problem for that? M: Twenty-one times three? T: What does twenty-one times three mean? M: Twenty-one, and take three out of twenty-one. T: Is that twenty-one times three or twenty-one divided by three? M: Twenty-one divided by three! T: Can you give me a multiplication problem? M: (Sits silently for over a minute). T: What are you doing? M: I am figuring out how many threes equal seven!

"I am figuring out how many threes equals seven" indicates to me that Maya was solving a new problem while she was sitting silently. She did not seem to use the units of three she made when counting backward by one as material for further operating. Why she did not take the result, seven, as how many units of three she had made, might be understood if one decenters and tries to assume Maya's point of view.

2.2 *Maya's Concepts of Multiplication and Division*

Maya had a result of counting, seven, that achieved her goal. To then take seven as a new given and formulate a multiplication problem would involve her becoming aware of how she had arrived at seven. She would have to re-present[3] the activity of counting and isolate the structure of the activity in its re-presented result. In other words, without any perceptual records of counting available, she would have to combine the units of three that she made and take seven as the given numerosity of the resulting unit containing those units of three. This seemed to demand an abstraction of the structure of operating that I hypothesized Maya had not made. To test this hypothesis, I investigated if she could additively combine six units of three and five units of three. To begin, I asked her to count out eighteen blocks by three.

T: Put eighteen blocks into that container. You can count them by three if you want. M: (Maya takes three blocks at a time and places them into the container). T: Give me a multiplication problem for that. M: (Long pause) three times six equals eighteen!

During the long pause, Maya apparently re-enacted counting to eighteen by three, silently recording how many times she made a unit of three – "1–2–3, that would be one", etc. She used the result of *this* activity to formulate her sentence "three times six equals eighteen" because she did not keep records of how many perceptual units of three blocks she made nor did she know that "3x6 = 18" from her regular classroom work. Continuing, I then asked Maya to put fifteen blocks into another container. Maya put them in by three and this time she kept track and said that she had five groups of three in the container. I then poured the contents of the two containers together and asked Maya to find the number of blocks in that combined collection of blocks using her units of three. The only answer Maya could give was "thirty" (Maya knew "five times six is thirty").

Her failure to additively combine the two lots of three is an excellent indicator of the nature of her concept of multiplication. She simply did not reason with units of three as she could reason with units of one. Her concept of multiplication consisted in anticipating[4] using a unit of three to segment counting by one; and actually segmenting counting by one into trios a given number of times constituted her multiplying actions. This concept of multiplication is less powerful than repeated addition because in that case the operations used involve reasoning with *composite* units rather than with units of one.

Maya's language – "take three out of twenty-one" – indicates that her concept of division did involve extracting three from 21 – a part-whole operation. But she did not seem to understand that the results of dividing would be a unit partitioned into units of three. Rather, her concept of division consisted in anticipating segmenting her number sequence from 21 down to and including one into trios. She could anticipate making units of three but did not *see* the abstract structure of the result as do children who have constructed quotative division. Actually segmenting counting by one into trios, and counting the trios so formed, constituted her dividing actions.

I have called operative schemes, like Maya's multiplying and dividing schemes, **mathematics of children** (Steffe, 1988a). While Maya's schemes have epistemological significance in the study of the genesis of multiplicative structures (cf. Piaget, 1970, pp. 13 ff), they also have educational significance for mathematics education. *As mathematics educators, we have a choice between using mathematics of children or conventional school mathematics as the basis on which to teach mathematics. Choosing the former is a fundamental requirement of constructivism for mathematics education.*

III. MATHEMATICS *FOR* CHILDREN

The first situation I illustrated to establish Maya's level of development in multiplication and division was not a problem for her. She didn't have a problem if that means "to search consciously for some action appropriate to attain a clearly conceived, but not immediately attainable, aim" (Polya, 1962). In Maya's assimilation of the division situation I posed, she apparently experienced an awareness of making more than one unit of three – an awareness of indefinite numerosity. This led to her making units containing three backward counting acts and coordinating those units with her standard number word sequence to make her awareness of indefinite numerosity definite. Counting seemed to be a constitutive part of her division concept and was activated by her awareness of indefinite numerosity. In this sense, the response of her dividing scheme was already available and was not an object of a search. This is reminiscent of a comment made by Piaget (1964) in a critique of associationism: "A stimulus is a stimulus only to the extent that it is significant and it becomes significant only to the extent that there is a structure which permits its assimilation, a structure which can integrate this stimulus but which at the same time sets off the response" (p. 15).

In scheme theory, what it means to have a problem is, first, for a child to have a scheme to which the given situation can be assimilated and, second, for the scheme's outcome to be inadequate to remove disequilibria that may have been created as a result of the assimiliation. If the child has not associated the action appropriate to remove a disequilibrium with the assimilated situation that triggered the scheme, the child's search for the action would be unsuccessful. This is precisely what happened when Maya tried to solve two of the tasks that I presented above. She seemed to have a problem when I gave her a multiplication task after she had used her division scheme and said "seven". Moreover, she seemed to have a problem when I combined into one container the six groups and the five groups of blocks she had made, and asked her how many trios were in the container. To solve either of these two problems would have involved a reorganization of her multiplying and dividing schemes.

At the time, I searched very hard for situations in which Maya would make these reorganizations. My searches for actions to solve the problem *I had* were as unsuccessful as Maya's searches for actions to solve the problems *she had.* In retrospect, I could find no reason to include inversion between multiplication and division and distributive reasoning[5] in Maya's zone of potential development[6] (Vygotsky, 1956) at any time during her third grade in school. *As mathematics educators, we have a choice between determining the mathematics* ***for*** *children through interactive communication or, on the other hand, through the conventional meaning of terms like "borrowing", "quotative division", or "distributive property". Choosing the former is a second requirement of constructivism for mathematics education.*

IV. TYPES OF LEARNING

The mathematics that children *can* learn is what I mean by mathematics for children. In the illustrations of mathematics that Maya could learn in the teaching experiment, I take learning as consisting in her modifications of the multiplying and dividing scheme explained above. For Piaget (1980b), assimilation is the fundamental relation involved in learning. It can involve or lead to perturbations, but the restoration of equilibrium that occurs as a result of neutralizing a perturbation by using a scheme might not involve learning. For example, after Maya had reached "seven" when counting trios of backward counting acts, she experienced a result that made definite her indefinite awareness of "more than one" that drove counting. She had used her counting scheme, but there were no modifications in it that were observable to me and, hence, no learning. On the other hand, modifications of a scheme would not occur without perturbations.

Functional accommodations. I have isolated two basic types of accommodation – functional and metamorphic (Steffe, 1988b). A functional accommodation of a scheme is any modification of the scheme that occurs in the context of using it. A child can make a functional accommodation in a scheme by **hindsight** (Hartmann, 1942) – by viewing the results of using a scheme in a novel way. For example, given how Maya used her dividing scheme, learning the basic division facts was in her zone of potential development. An association between a division situation such as Maya experienced and the result of using the scheme could have been established by Maya had I encouraged her to reflect on her immediate past experience and to abstract the involved language ("twenty-one divided by three" and "seven"). This is reminiscent of the principle of association explained by Guthrie (1942): "A stimulus pattern that is acting at the time of a response will, if it recurs, tend to produce the response" (p. 23). Although I disagree with him that the principle of association is the basic event in learning, its reformulation in scheme theory is powerful enough to explain how children like Maya might form associations called "basic facts" as a result of what Piaget calls reflective abstraction (Piaget, 1980a).

V. PROCEDURAL FUNCTIONAL ACCOMMODATIONS

Functional accommodations that I call procedural are not restricted to hindsight. A child might also modify the activity of a scheme without curtailing it. For example, approximately one year after I had isolated Maya's schemes for multiplying and dividing, I again investigated if distributive reasoning was in her zone of potential development. This investigation was promoted by the progress she had made in her concept of multiplication. It had become an iterative concept – "seven times twelve" now meant twelve seven times. So, it seemed to be plausible that she could learn to combine twelve four times and twelve three times, to produce twelve seven times, and to separate the latter to produce the former. She did modify the activity of iterating a composite unit in the following protocol.

T: (Places a strip of adding machine paper in front of Maya) this is 36 inches long. How many 12-inch rulers is it going to take to measure it? M: Three. T: How did you know? Did you know that three times twelve is thirty-six? M: (Shakes her head "yes"). T: (The 12-inch ruler was divided into a 10-inch and a 2-inch ruler) now, if you put the tens down first, how many of those would it take and how many of the twos? M: It would take three of these (the

10-inch ruler) and three of these (the 2-inch ruler). T: Tell me why. M: Because 3 tens is thirty. Two times three is six and thirty plus six would be thirty-six!

As her teacher, I was excited because it seemed possible that Maya had deeply reorganized her multiplication concept. If this proved to be the case, then Maya would have a solid basis for constructing standard multiplying algorithms as **child-generated algorithms** (cf. Steffe, 1983; Hatfield, 1976) because she would be able to **produce** these algorithms as functional modifications (accommodations) of her current operative multiplying scheme.

Another possibility was that Maya had merely modified the **activity of multiplying rather than her concept of multiplication**. We do see in the protocol that Maya explained why it would take three of each smaller ruler by finding each product and adding their results together. There didn't seem to be the logical necessity that an understanding of distributivity would provide – there would be no reason to find the sub-products and add them together if distributivity was a property of her concept. However, she might have used distributivity to obtain the result and then used computing to **explain why it worked.**

A third possibility was that Maya had made a local modification in her multiplication concept as well as a modification in the activity of muliplying because, in the problem preceding the one of the protocol, I had presented her with a 24 inch strip of adding machine paper and the separated 12-inch ruler which, in turn, separated the adding machine paper into two congruent parts. Without moving the ruler, Maya immediately saw that it would take two 12-inch rulers and two of each smaller ruler to cover the strip of paper. She could easily anticipate putting another 12-inch ruler beside the one on the paper to cover the uncovered part, closing the potential measurement activity. The results of measuring thus formed a visual whole and the parts could be easily isolated and recombined **with respect to the visual whole.**

If the third interpretation proved to be the most plausible, the local modification of her multiplication concept would be restricted to situations similar to the one in which it occurred. To investigate if this was so, I presented her with an 84-inch piece of adding machine paper.

T: See if you can do this one – 84 inches (places an 84-inch strip of adding machine paper in front of Maya). Suppose you use the 10-inch one first and then the 2-inch one second. They have to be equal like they were before. How many of each kind would it take? M: (Long pause) do they have to be equal? T: Tell me one that isn't. M: I don't have one. T: If you find one, tell me. M: You could use eight of these (10-inch ruler) and two of these (2-inch ruler). T: Can you find two that are equal? M: (Long pause where Maya is in deep concentration). I can't find any!

It would have required a deep insight for Maya to use her knowledge of the associated multiplication fact. Had she said "seven because seven times twelve is eighty-four", I would have inferred that she understood that so many iterations of a 12-inch ruler is equivalent to the same number of iterations of a 10-inch and a 2-inch ruler. From this episode, I concluded that the third interpretation was the most plausible.

However, I still was not satisfied. Could this local modification lead to a deeper reorganization of her multiplication concept? So, I went on, encouraging Maya to experiment. She tried 9 upon my suggestion and found that it did not work because "nine times ten is ninety". She then selected 7 as a possibility after I asked her what the next logical choice would be. After finding the two sub-products and adding their results, I asked her to tell me the result of multiplying seven and twelve.

T: By the way, what is seven times twelve? M: (Immediately) eighty-four. T: How did you know that ? Did you just know it or did you use this to help you (pointing to the measurement situation)? M: No (indicating she did not use her measurement results). T: Could you use this to help you? Suppose you did not know seven times twelve, how could you use it? M: Seven twelves, you would put those two together (the 10-inch and the 2-inch ruler).

Maya clearly understood that if she placed the 12-inch ruler down seven times, it would measure 84 inches, just as would placing the 10-inch *and* the 2-inch ruler down seven times. However, she combined the 10-inch and the 2-inch rulers together to make the 12-inch ruler as an **explanation** for why the basic fact and the measurement led to the same result. Distributivity was still not a property of her reasoning **before measuring.**

I do believe that using products for which Maya had no established associative bonds held promise as a learning environment in which she could reorganize her concept of multiplication, because strategies of reasoning like the one displayed could be encouraged. Distributivity was now in her zone of potential development. But, I still could not infer that she had made a general reorganization of her concept of multiplication. She did modify her activity of multiplying in the measurement situation and there was also a local modification of her multiplication concept. But there was no general reorganization of muliplication that would be manifest in the flexibility made possible by distributive reasoning.

I would not have known this had I not generated possible interpretations (von Glasersfeld, 1987) of Maya's modification of multiplying and tested which was the most plausible by investigating the generality of the modification. In the situation I used as a test, I simply varied the length of the strip of paper with the intention of creating a problem that Maya could solve with my help. Our interactive communication finally enabled me to interpret her modification. *As mathematics educators, we have a choice between interpreting children's mathematical activity in learning environments through interactive communication, making qualitative distinctions, or focusing on the result or product of their activity. Choosing the former is a third requirement of constructivism for mathematics education.*

VI. LACK OF INSIGHT

In the case of the 36-inch piece of adding machine paper, Maya did not have a problem. Her utterance "three" indicated an assimilating generalization – differences in the new situation were manifest in the results of using the operations of the assimilating scheme. Her initial response in the case of the 84-inch strip was also the result of assimilating generalization. But, when I said "find two that are equal", she genuinely had a problem whose immediate solution would involve an insight as explained by Wertheimer (1959).

> what a change, when ... all the parts suddenly form a consistent clear whole, in a new orientation, in strong reorganization and recentering, all fitting the structural requirements (p. 57).

If an insight had led to the distributive actions that solved the problem when I asked her to find two that were equal, it would be a midsight[7] (Gates, 1942). But, given Maya's current iterative multiplicative concept, there was little chance of her producing such an insight as a midsight, foresight, or hindsight. The modification in her multiplication concept was restricted to interpreting the multiplicand as so many tens and so many ones.

Other than her iterative multiplicative concept and the functional accommodations explained above, Maya made few modifications in the primitive multiplicative scheme I isolated while she was in the third grade. However, as a result of her regular classroom instruction, she had learned the multiplication facts and the paper and pencil computational algorithms for multiplying one and two digit numbers as explained by Thorndike (1922) more than 65 years ago. Her algorithms consisted of an organization of associative bonds that included the basic facts, carrying, etc. (Gates, 1942). Maya had not produced the computational algorithms as modifications of operative schemes, if for no other reason than that she lacked distributivity. Her computational algorithms were connected to, rather than contained in, her iterative multiplicative concept. As such, Maya had "learned" a new scheme for multiplying that was not a modification of her more primitive scheme. It is little wonder that Nesher (1986) has commented, "No one has succeeded in demonstrating that *understanding* improves algorithmic performance" (p. 16), because algorithms are viewed as procedures disembodied from the mathematics *of* children and are imposed upon them without regard to the nature of their schemes. *As mathematics educators, we have a choice between taking assimilation as the fundamental relation involved in learning and learning as consisting in the modifications of schemes or, instead, taking association as the fundamental relation involved in learning and learning as consisting of establishing and organizing bonds into a system. Choosing the former is a fourth requirement of constructivism for mathematics education.* In that case where the associations are established as a result of reflective abstraction, one can include association as a relation involved in learning – but not the fundamental one – because the associations can function in place of the curtailed operations.

Metamorphic accommodation. Had Maya abstracted the structure of her distributive actions after finding that "seven" worked, it would have amounted to a reorganization of her concept as explained by Wertheimer even though it would not have been a foresight. Regardless of when it occurs in a problem solving episode, such an insight would be an example of a metamorphic accommodation because it would lead to a general reorganization of the involved concept. From the observer's perspective, if Maya spontaneously used distributive reasoning as a strategy to rename products, then a general reorganization could be inferred.

Any accommodation that occurs independently and involves a general reorganization of a scheme is called "metamorphic accommodation". Had Maya's multiplication concept been an abstract concept on a par with the concept of quotative division as I have explained it, she could have produced distributive reasoning as an insight when she solved some problem. However, given her iterative concept, an interiorization of the iterative actions would have been necessary for her to substantially reorganize the concept. Interiorization is the most general form of abstraction.

> It leads to the isolation of structure (form), pattern (coordination), and operations (actions) from experiential things and activities; an interiorized entity is purged of its sensory-motor material (In Steffe, Cobb & von Glasersfeld, 1988, p. 337).

Knowing what interiorization leads to does not explain its process. I have found the process to involve the use of operations in the recreation of an immediate past or possible experience (Steffe, 1988b). For example, when Maya explained, "Because 3 tens is thirty. Two times three is six and thirty plus six would be thirty-six!" I believe she used numerical operations to enact her possible measuring experience. These numerical operations are precisely the operations that are needed to purge measuring activity of its sensory-motor material. However, one or more such episodes was insufficient to complete a developmental metamorphic accommodation – a metamorphic accommodation that occurs as a global **result** of such specific experiential episodes (Steffe, 1988b). It involves interiorization and reorganization of specific operations, where the reorganization is manifest in independent use of the operations in solving problems that have, from the researcher's perspective, analogous structure.

From the researcher's perspective, the purpose for engaging children in goal-directed activity that includes problem solving is not simply the solution of specific problems. The primary reason is to encourage the interiorization and reorganization of the involved schemes as a result of the activity. These interiorized and reorganized schemes constitute operative mathematical concepts that are constructed by means of reflective abstraction. *As mathematics educators, we can treat operative mathematical concepts in one of two ways: (1) as constructed by children as a result of goal-directed activity in learning environments, or (2) a priori to mathematics learning and disembodied from learning – as what the children are to learn this school year, this week, or today regardless of whether or how the children might learn them. Choosing the former is a fifth requirement of constructivism for mathematics education.*

VII. CHILDREN'S MATHEMATICAL KNOWLEDGE AS TEACHER KNOWLEDGE

Maya's schemes were formulated in my attempts to understand her ways and means of solving the situations I posed. Specifying children's schemes does involve creating and posing situations that might reveal those schemes from their sensory-motor to their operative forms. But I do not focus on the situations per se. Rather, I focus on children's solution attempts in a search for the schemes they might use in assimilation and solution. If I am able to specify the schemes that explain what children do, I can then create and pose new situations with the intention of isolating any modification they might make. Using this kind of a strategy, it is possible to learn the mathematical knowledge of the children I am charged with teaching.

It is not my intention to argue that my explanation of Maya's operative schemes and their modifications are identical to Maya's knowledge in this content area (cf. von Glasersfeld, 1983). Rather, my argument is that Maya's language and actions were the experiences available to me in a learning environment as I strove to learn her schemes of operating and that my interpretation could be made only in terms of my own knowledge. So, from my point of view, my description of Maya's mathematical knowledge is unavoidably an expression of my own concepts and operations. Although I do acknowledge that Maya had a mathematical reality separate from my own, I had no direct access to it.

Childen's mathematical knowledge, then, refers to our models of the mathematical knowledge of children. Learning children's mathematical knowledge is part of what it means to be a mathematics teacher and no one else can assume this responsibility for the teacher. *As mathematics educators, we can take the responsibility for learning children's mathematical knowledge or we can take it as a given, similar to other parts of our own mathematical knowledge. Choosing the former is the sixth requirement of constructivism for mathematics education.*

VIII. TEACHING AS GOAL DIRECTED ACTIVITY

A relativistic world view legitimatizes children's mathematical knowledge. To view students as important others who have a mathematics that is indeed worth knowing represents a general change of paradigm. The focus is shifted from the teacher, not to the students, but to the interactive communication that transpires between teachers and students, among students, or among teachers (von Foerster, 1984; von Glasersfeld, 1988; Bauersfeld, 1988). Interactive communication constitutes the central activity of teaching and this activity is goal-directed in a way that is analogous to children's mathematical activity as I explained it. *Mathematics educators may view mathematics teaching in this way or as the transmission of knowledge (McKnight, et al. 1987). Making the first choice is the seventh requirement of constructivism for mathematics education.*

The nonrelativistic view of teaching as the transmission of knowledge is based on the idea that students must come to know mathematics as it is, a view that ignores the mathematical reality of children. It is a "missing child" paradigm and I wonder whether its advocates would consider the mathematics of children as what it is, not what the teachers wish it to be, as they do abstract mathematics. Constructivist teachers do not deny that children have mathematical realities distinct from their own. To the contrary, they actively seek to learn and to modify those realities. The teachers' models are not simply as they might wish them to be, for they are always subject to modification in interactive communication. What children do constrains teachers and constitutes a critical point of contact with the actual mathematical knowledge of children.

IX. LEARNING ENVIRONMENTS

The result of an assimilation of a particular situation is an experience of the situation and this experience constitutes a learning environment in the immediate here and now. From a teacher's perspective, a child's learning environment consists of what the teacher intends for the child to learn after the child assimilates a particular situation. To distinguish between the teacher's perspective of a child's learning environment and the child's perspective, "possible learning environment" and simply "learning environment" are used to refer to the teacher's frame of reference and to the child's frame of reference, respectively. Possible learning environments are determined by a teacher in part with respect to children's zones of potential development (Vygotsky, 1956) and in part with respect to children's actual level of development (Sinclair, 1988). The latter is determined by the situations the children can independently solve and the former is determined by the problems the children can solve with the help of a teacher. In the constructivist view, a zone of potential development relative to a child's specific scheme is determined by the modifications of the scheme the child might make in, or as a result of, interactive communication in learning environments. Like possible learning environments, a zone of potential development is the teacher's concept.

Because zones of potential development and possible learning environments depend on each other, it is possible for a child to have differing zones of potential development with respect to the same scheme. The particular modifications of a scheme could diverge in one of several directions depending on the possible learning environments encountered by the child which, in turn, are dependent on particular modifications. This realization places the teacher in a crucial position with respect to the mathematical development of his or her students. Nevertheless, the unavoidable dependence does not mean that a child's zone of potential development is wholly dependent on possible learning environments, nor does it mean that the child can learn anything the teacher wants the child to learn in the immediate future.

A particular modification of a scheme cannot be caused by a teacher any more than nutriments can cause plants to grow – nutriments are used by the plants for growth but they do not cause plant growth. Teachers are constrained in specifying what they place in a child's zone of potential development by what the child makes from their experiences in particular learning environments. Zones of potential development are negotiated, then, through the interactive communication that transpires in learning environments.

As a concept, a learning environment can be thought of as a variable experiential field whose contents are specified by its participants. *As mathematics educators, we have a choice between viewing learning environments in that way or taking the situations of learning as being a priori and invariant throughout their use by children – considering our frame of reference as the only one. Choosing the former is an eighth requirement of constructivism for mathematics education.*

X. TEACHING TO LEARN

Piaget (1964) identified experience as one crucial factor of intellectual development and this applies to us as mathematics educators as well as to our students. Mathematics educators who aim at learning children's mathematical knowledge and its evolution must assume that there will be some regularity in their experiential world. These recurring patterns are the basis for forming concepts (or models) of the developing mathematical knowledge of children. In fact, the term "model" refers to a conceptual generalization that is abstracted from a group of experiences for the purpose of categorizing and systematizing new experiences (von Glasersfeld & Steffe, 1987). I believe that mathematics educators can reach a point where they may feel justified in making predictions about further experiences of the child or about the experiences of children like those with whom they are currently working.

The generative power of children can be impressive when they are working in learning environments that are conducive to constructive activity (Steffe, Cobb, & von Glasersfeld, 1988). However, children's mathematical knowledge has only begun to be charted. *Mathematics educators at all levels have an exciting choice between being participants in specifying the generative mathematical power of children or taking what children can learn as being already specified by an a priori curriculum. Choosing the former is a ninth requirement of constructivism for mathematics education.*

For mathematics educators who opt for constructivism, the following ten principal goals might serve as a guide in their research activities.

1. To learn how to communicate mathematically with students.
2. To learn how to engage students in goal-directed mathematical activity.
3. To learn the mathematics of the students they teach.
4. To learn how to organize possible mathematical environments.
5. To learn the content of variable experiential fields – the mathematical experience of students.
6. To learn the mathematics *for* the students they teach.
7. To learn how to foster reflection and abstraction in the context of goal-directed mathematical activity.
8. To learn how to encourage students to communicate mathematically among themselves.
9. To learn how to foster student motivation and how to sustain learning over a long period of time.
10. To learn how to communicate pedagogically as well as mathematically with other mathematics educators.

Encouraging reflection and abstraction in learning environments can be very important. I agree with Treffers (1987) that students should be given the opportunity to reflect on their own mathematical experiences by being asked a critical question or by exploration in a novel but related context. In each case, Treffers assumes that at least some successful mathematical learning has occurred so there is something to reflect on – something to be analyzed. Becoming aware of one's own ways and means of operating when modifying them to meet certain constraints is precisely what is involved in reflection and abstraction and leads to interiorization of the involved operations. In fact, one of my explicit goals when teaching children is for them to become increasingly aware of their own developing mathematical power and autonomy in controlling and monitoring their learning activity. I believe this developing sense of the self doing mathematics successfully and independently is crucial in developing student motivation and in sustaining learning over a period of time. Children need to take responsibility for their own learning as early as possible in their life and they need to distinguish their own conceptual operations and experience from that of others.

We should not confuse interactive communication with mathematical activity. However, the latter can be greatly influenced by the former and this is why I believe that it is crucial for teachers to learn to communicate mathematically with children and how to encourage children to communicate mathematically among themselves (Cobb, 1987; Wood & Yackel, 1990). It is essential that the student's construction of mathematical reality be a part of their social construction of reality, because it is through the processes involved in socialization that personal mathematical knowledge can be contrasted with the mathematical knowledge of others; and this comparison is the likeliest way to make progress.

XI. FINAL COMMENTS

A goal structure for mathematics education such as the one elaborated by Treffers (1987) is needed in specifying possible learning environments by teachers. But this element of possible learning environments is just as dependent on the experiential fields that constitute learning environments as the latter are dependent on the former. Mathematics educators should not take their goals for mathematics education as fixed ideals that stand uninfluenced by their teaching experiences. Goal structures that are established prior to experience are only starting points and must undergo experiential transformations in actual learning and teaching episodes. Mathematics educators can have a goal structure that initially identifies mathematics for children, yet what that mathematics will eventually consist of is not simply a transformation of the initial goal structure. Hypotheses can be formulated and tested, but mathematics educators should expect to learn certain elements of children's mathematical knowledge only through their own experiential abstraction. These abstracted goal structures should be modified and made public through interactive communication among mathematics educators for the same reason that mathematical schemes should be modified and made public among children.

As a sociological construct, a possible learning environment is not simply a collection of situations that might be presented to children. It contains a goal structure pertaining to the mathematical knowledge of the students it is intended for, a possible sequence of representative situations the teacher can pose to the students, critical questions that might be asked by the teacher, a possible itinerary of modifications of the involved schemes, and sample interactive communication that should be encouraged among the students. As such, a possible learning environment is a conceptual generalization a teacher can use in the creation of learning environments with his or her students. A learning environment is literally created by its participants and a teacher can use his or her conceptual generalization in this context to guide and interpret children's mathematical activity. But, as with any other goal-directed activity, the results of the activity can modify the guiding concepts. In other words, a general goal of mathematics teaching is for teachers as well as students to learn, and the primary goal of the constructivist teaching experiment is but a microcosm of this general goal of mathematics teaching.

NOTES

1. The research on which this paper was based was supported by the National Science Foundation under Grant No. MDR-8550463 and the Institute of Behavioral Research of the University of Georgia. All opinions and findings are those of the author and are not necessarily representative of the sponsoring agencies.
2. An **assimilation** is the integration of new objects or new situations and events into previous schemes (Piaget, 1980b, p. 164).

3. A **re-presentation** is the re-creation of a perceptual or motor experience without actual sensory-motor material; one is in it and, in a very real sense, one acts again. It is like a playback.
4. A **scheme is anticipatory** when recognition of an experiential situation leads to a re-presentation of the associated material (or mental) actions and/or result.
5. **Distributive reasoning** can occur in various forms. For example, if a child finds how many candies in five packages with 12 candies in each package by finding the sum of thirty and thirty, that would be an indication of distributive reasoning.
6. **In constructivism**, a zone of potential development of a specific mathematical concept is determined by the modifications of the concept the student might make in, or as a result of, interactive communication in a mathematical learning environment.
7. A **midsight** is an insight that occurs during the activity of solving a problem.

REFERENCES

Bauersfeld, H. (1988). "Interaction, Construction, and Knowledge: Alternative Perspectives for Mathematics Education". In D. Grouws, & T. Cooney (eds.), *Perspectives on Effective Mathematics Teaching*, Reston: National Council of Teachers of Mathematics, 27–46.

Cobb, P. (1987). "Toward A Reconciliation of the Individual and the Social". Manuscript copy, Purdue University.

Cobb, P. & L.P. Steffe, (1983). "The Constructivist Researcher as Teacher and Model Builder." *Journal for Research in Mathematics Education*, 14(2), 83–94.

Gates, A.I. (1942). "Connectionism: Present Concepts and Interpretations". In N.B. Henry (ed.), *The Psychology of Learning: Forty-First Yearbook of the NSSE* (pp. 141–64). Chicago: University of Chicago Press.

Guthrie, E.R. (1942). "Conditioning: A Theory of Learning in Terms of Stimulus, Response, and Association". In N.B. Henry (ed.), *The Psychology Learning: Forty-First Yearbook of the NSSE* (pp. 17–60). Chicago: University of Chicago Press.

Hartmann, G.W. (1942). "The Field Theory of Learning and Its Educational Consequences". In N.B. Henry (ed.), *The Psychology of Learning: Forty-First Yearbook of the NSSE* (pp. 165–214). Chicago: University of Chicago Press.

Hatfield, L. (1976). "Child-Generated Computational Algorithms: An Exploratory Investigation of Cognitive Development and Mathematics Instruction With Grade II Children", unpublished manuscript, Athens: University of Georgia.

Hawkins, D. (1973). "Nature, Man and Mathematics". In A.G. Howson, (ed.), *Developments in Mathematics Education*, 115–35. Cambridge: Cambridge University Press.

Hunting, R.P. (1983). "Emerging Methodologies for Understanding Internal Processes Governing Children's Mathematical Behavior". *The Australian Journal of Education*, 27(1), 45–61.

McKnight, C. (1987). *The Underachieving Curriculum*. Champaign: Stipes Publishing Company.

Nesher, P. (1986). "Are Mathematical Understanding and Algorithmic Performance Related?" Invited Address for the 1986 Annual Meeting of the American Educational Research Association, Division C–4, San Francisco.

Piaget, J. (1964). "Development and Learning". In R.E. Ripple & V.N. Rockcastle (eds.), *Piaget Rediscovered: Report of the Conference on Cognitive Studies and Curriculum Development, 7–20*. Ithaca: Cornell University.

— (1970). *Genetic Epistemology*. NY: Columbia University Press.

— (1980a). "The Psychogenesis of Knowledge and Its Epistemological Significance". In M. Piattelli-Palmarini, (ed.), *Language and Learning: The Debate Between Jean Piaget and*

Noam Chomsky (pp. 23–34). Cambridge: Harvard University Press.

— (1980b). "Schemes of Action and Language Learning". In M. Piattelli-Palmarini (ed.), *Language and Learning: The Debate Between Jean Piaget and Noam Chomsky* (pp. 23–34). Cambridge: Harvard University Press.

Polya, G. (1962). *Mathematical Discovery: On Understanding, Learning and Teaching Problem Solving*. NY: John Wiley & Sons, Inc.

Sinclair, H. (1988). "Foreword". In L.P. Steffe & T. Wood (eds.), *Transforming Early Childhood Mathematics Education*: Hillsdale: Laurence Erlbaum.

— (1990). "Learning: The Interactive Recreation of Knowledge". In L.P. Steffe and T. Wood (eds.), *Transforming Early Childhood Mathematics Education*, 19–29: Hillsdale: Laurence Erlbaum.

Steffe, L.P. (1983). "Children's Algorithms As Schemes". *Educational Studies in Mathematics*. 14, 109–25.

— (1984). "The Teaching Experiment Methodology in a Constructivist Research Program". In M. Zweng (ed.), *Proceedings of the Fourth International Congress on Mathematical Education* (pp. 469–71), Boston: Birkhauser.

— (1988a). "Children's Construction of Number Sequences and Multiplying Schemes". In J. Hiebert & M. Behr (eds.), *Number Concepts in the Middle Grades* (pp. 119–46). Hillsdale: Lawrence Erlbaum Associates.

— (1988b). "Modifications of the Counting Scheme". In L.P. Steffe, P. Cobb, (with E. von Glasersfeld), *Construction of Arithmetical Meanings and Strategies*, 284–322. NY: Springer-Verlag.

Steffe, L.P & E. von Glasersfeld (1985). "Child Generated Multiplying and Dividing Algorithms: A Teaching Experiment". NSF Grant No. MDR–8550463.

Steffe, L.P. & P. Cobb (with E. von Glasersfeld) (1988). *Construction of Arithmetical Meanings and Strategies*. NY: Springer-Verlag.

Thorndike, E. (1922). *The Psychology of Arithmetic*. NY: MacMillan.

Treffers, A. (1987). *Three Dimensions: A Model of Goal and Theory Description in Mathematics Instruction – The Wiskobas Project*. Boston: D. Reidel Publishing Company.

von Foerster, H. (1984). "On Constructing A Reality". In P. Watzlawick (ed.), *The Invented Reality: How Do We Know What We Believe We Know*? 41–61. London: W.W. Norton & Company.

von Glasersfeld, E. (1983). "Learning As Constructive Activity". In J.C. Bergeron & N. Herscovics (eds.), *Proceedings of the Fifth Annual Meeting of PME-NA*, 41–69. Montreal, Canada.

— (1987). "Feedback, Induction, and Epistemology". In E. von Glasersfeld, *The Construction of Knowledge* (pp. 219–30). Seaside, CA: Intersystems Publications.

— (1990). "Environment and Communication". In L.P. Steffe & T. Wood (eds.), *Transforming Early Childhood Mathematics Education*, 30–38. Hillsdale: Laurence Erlbaum.

von Glasersfeld E. & L.P. Steffe (In Press). "Conceptual Models in Educational Research and Practice". *Journal of Educational Thought*, 25(2).

Vygotsky, L.S. (1956). "Learning and Mental Development at School Age". In A.N. Leontiev & A.R. Luria (eds.), *Selected Psychological Works*. Moscow, 383–452 (Article written in 1934).

Wertheimer, M. (1959). *Productive Thinking*. NY: Harper & Row.

Wood, T. & E. Yackel (1990). "Teacher's Role in the Development of Collaborative Dialogue Within Small Group Interactions". In L.P. Steffe & T. Wood (eds.), *Transforming Early Childhood Mathematics Education*, 244–252. Hillsdale: Laurence Erlbaum.

DIDACTIC CONSTRUCTIVISM

JAN VAN DEN BRINK

State University of Utrecht, The Netherlands

I. INTRODUCTION

This chapter deals with the problems of operationalizing Radical Constructivism (hereafter referred to as RC) in teaching. In that context we call it 'didactic constructivism'.

1.1 Can RC be applied at all?

One wonders if RC can be applied in teaching. It is certainly incorrect to start by developing a general epistemology that is rooted in philosophy, psychology and cybernetics (see the Introduction by Von Glasersfeld) and then go on to seek proof of the validity of various constructivist principles in teaching practice and didactics.

Nevertheless we found amazing similarities between this epistemology and the mathematics education developed in The Netherlands in 1970 as part of the 'Wiskobas' project. It looked as if there had been two parallel developments but in different domains.

We will return to this matter later (1.4).

We feel that it is important, both for theory and teaching, to draw attention to this similarity.

As a result of the examples taken from teaching practice, the theory will become clearer and more accessible to teachers and it is conceivable that they in turn will approach their instruction in a new light.

1.2 Should RC be applied?

Another reason for applying constructivism in teaching is not only that it is possible, but also that it is needed. Often theoreticians disregard the consequences of their theories for teaching practice. They seem to think that theory will automatically be integrated into teaching in schools. There are however examples that demonstrate that sometimes this does not occur.

For instance, the conservation tests of Piaget were taken straight over into arithmetic teaching without being first examined from new points of view of teaching methodology. Cylinders with different diameters are now to be found

E. von Glasersfeld (ed.), Radical Constructivism in Mathematics Education, 195–227.

in every kindergarten. They are used to practice and test the concept of conservation. First of all, it should be pointed out that Piaget illustrated the concept conservation in tests connected with weight and time (1) and not solely in connection with contents of glasses; but the latter tests are better known. Therefore Piaget himself was not responsible for the restriction of conservation to the contents of glasses. But he was responsible, indeed, for the unfortunate fact that he gave no indications about how conservation fitted into the curriculum and about the consequences of it for teaching mathematics. In the Wiskobas project however the conservation principle was integrated into its new teaching methodology.

Example: Filling the vase
The teacher asks a pupil: 'Fill this vase half-full'.

Figure 1.

The children fill the 'flask' until the water is 'half way'.

However, that does not mean that the vase is half-full. But that does not bother the children. They have enough trouble keeping the water steady. Then the teacher points out that the flask is not half-full. The stopper is put in the flask. Not a drop more can be added or poured off. The children are extremely surprised when the flask is turned around: suddenly the flask is more than half-full!

"Magic water" exclaims a pupil. Another pupil tries to explain what has happened: 'There's a big air bubble over the water and it will stay there. The water will stay below it'.

But Renate can provide an explanation: 'This part is bigger' and she points to the bulb of the flask, 'higher up it's narrower'. If the flask is to be really half-full, should water be added or poured off? ...

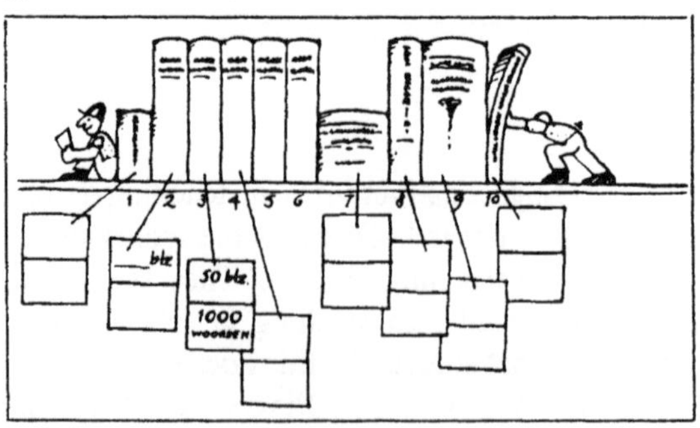

Figure 2.

It is clear that this example demonstrates better than the experimental set-up with different glasses how conservation relating to content can acquire a meaning in mathematics education. As educationalists we are not satisfied with the remark that the quantity of water does not change.

We seize upon this obvious fact in order to create a *conflict situation* which extends the child's concept of content in the direction we want: the appearance of the content *depends on various dimensions* (height, shape, length, breadth).

Look at the books and estimate the number of pages and words in each book via the given numbers of one book.

1.3 Ways of applying RC

RC can be applied in various ways. Just as with Piaget's tests, one can illustrate constructivist standpoints with striking examples from classroom teaching or with underlying didactical principles. An example. One of the basic principles of RC is that '(...) knowledge (...) concerns the construction of paths of actions and thinking that an unfathomable 'reality' leaves open for us to tread. The test of knowledge, is not whether or not it accurately *matches* the world as it might be 'in itself' (...) but whether or not it *fits* the pursuit of our goals (...) within the confines of our own experiential world (...)' (Von Glasersfeld, 1988, p. 2).

The following incident in the kindergarten is a nice illustration of the above. All the children (aged 5–6) had recently travelled on a city bus; it had stopped repeatedly at bus-stops. In the Wiskobas project the bus is an important model for introducing addition and subtraction. We re-created the bus situation in the classroom: one pupil is the driver and wears a cap; the other children, standing in a line behind him, are the passengers.

At the first stop all the children got off, which was not what the teacher expected. Some of the children thought they were on a school-trip and not on the city bus. Others thought they were on the city bus and were going to 'granny' but they failed to notice that the bus had stopped repeatedly. That did not matter to them on their real way with mother to see 'granny'. Others wondered whether the 'game' had finished at the first stop ---

Each child had his or her own particular idea which fitted into the play-situation of getting off. The surprise shown by the teacher however quickly put an end to the confusion. The children revised their ideas and quickly hopped back on to the bus.

It is instructionally interesting to see how we can consciously provoke the 'fitting' of children by changing the context or the situation (e.g., by play acting and by showing suprise). With the help of such instructional manoeuvres we can develop *meaningful* knowledge in children, in the sense of knowledge that *fits* into a situation as envisaged by RC (see 3.3). I should be able to explain important principles of RC by taking examples from the classroom and by means of educational manoeuvres. But this would be unsatisfactory from an educational point of view. We must try to discover whether RC can make a contribution to teaching methodology and can help us to establish educational principles and develop striking examples. Should there not perhaps be a kind of 'grammar' which could guide us in our search and which is based on the principles of radical constructivism?

Let us explain this idea more fully. There is a general educational problem: children make thought-constructs; but for the teacher, the questions are:

- o How do I get the children to make thought – constructs?
- o How can I get the children to develop their knowledge of the world in the direction I want it to develop?

In the classroom these questions are answered in various ways.

- o In *traditional arithmetic* lessons tasks are *demonstrated* and then *initiated.* The answer to the questions is: by letting the children practise as much as possible.
- o It is only fairly recently that children in school have been required *to devise tasks* themselves and to discuss them. However, letting children develop their own constructs is not always meaningful (or fitting) in a children's context.
- o An even more recent development is to find contexts within which it is meaningful for children to design their own constructs. A good example is allowing children to compile an arithmetic book (Van den Brink, 1987)

Example: children as authors of an arithmetic book.

In the middle of the school year it was suggested that children of grade 1 should write an arithmetic book for the children who would be in that grade in the following year. One page was written per month; each pupil wrote 6 pages in all; the pages contained tasks for the beginning of the school year, arithmetical games, tasks for the weeks before Christmas, practice sums, etc.

The results were amazing.

- *There were hardly any mistakes in the books.*
- *The numbers used became gradually larger.*
- *All kinds of out-of-school arithmetical games were mentioned.*
- *New practice methods were devised.*

In summary, the design of an arithmetic book for others, was a meaningful context for children in which their own constructs were fitting.

- o And going a step further: Are there principles and ideas on the basis of which we can find similar contexts which make arithmetic meaningful for children?

RC provides hints (think for instance of the above mentioned examples concerning the difference between 'matching' and 'fitting') about various ways in which teaching-methodology can be enriched.

1.4 The Wiskobas project – or in other words the realistic movement in mathematics education in The Netherlands

The Wiskobas project introduced a realistic stream into mathematics teaching in The Netherlands and led to major changes in the arithmetic textbooks. There is a certain correlation between this stream in mathematics education and RC which is not coincidental. RC is influenced by the intuitionism of Brouwer. The realistic stream was instigated by Freudenthal, who was a student of Brouwer. The fundamental idea of didactic realism (hereafter referred to as DR) is that mathematics and arithmetic are regarded as a human activity rather than as a series of problems and systems. Central to DR is the notion of *realization*. By that we mean that the pupil creates a conceptual representation (makes a 'construct') in connection with a context, a task, a formula, a sum, an action. The pupil makes a construct, because otherwise he would not understand the task, the formula, etc. Constructing in the sense of making conceptual representations is therefore an essential feature of arithmetic and mathematics teaching in The Netherlands.

The links between RC and DR are clear, although the term didactic *realism* may cause confusion. The term is derived from the verb 'to realize' (to imagine, etc.) and not from the use of contexts from a so-called objective reality outside the pupil.

Von Glasersfeld (1988, p. 5) writes: '(...) While realists tend to think that their models should (...) reflect the students as they *really* are, constructivists must remain aware of the fact that models cannot reflect anything but the model builder's own conceptual constructs (...)'. This holds also for didactic realism which has the verb to realize as its central core (2).

One could say that 'to realize' means almost the same as 'to construct'. Both means: to think up something and make something. Fundamental to both concepts is that the task of mathematics teaching is to investigate '(...) how the students could be given the opportunity to modify their structures so that they lead to mathematical actions that might be considered compatible with the instructor's expectations and goals (...)' (Von Glasersfeld, 1988, p.6).

Therefore there should not be any difference between RC and DR. But teaching has forced DR to make a closer study of how one should organize and achieve the above in the classroom setting. Didactic realists could not be satisfied with purely epistemological considerations; they had to design a whole teaching program (the Wiskobas integration plan) which would provide the educational means to lead the children to make conceptual constructs.

In our interpretation the verb 'to realize' embraces two activities: one is '*to construct*', the other is '*to confront*'. You create a conceptual representation. But if you then confront this with the ideas of others or with the execution of the idea you 'realize' your first thought again and discover mistakes in it which you did not perceive before.

Therefore the concept of realization contains two components: a 'constructing' one and a 'confronting' one. All kinds of practical aids have been devised to stimulate the construction of thoughts in children. The aids are all based on confrontations, eg. with invisible quantities, a shortage of material, changes in the circumstances in which the task is set, etc.

We will take a closer look at didactic realism in the following section.

II. EXAMPLES OF DIDACTIC REALISM IN THE RADICAL CONSTRUCTIVISTIC FRAMEWORK

2.1 Three guiding principles of the DR

It should be clear from the foregoing that DR points to at least three ways in which children can be stimulated to construct arithmetical concepts (3):

1. Thought constructs should be *bound to contexts* and to expressive activities and expressive arithmetical languages.
2. There should be *confrontation* between various constructs.
3. Constructions should be meaningful and *senseful* (*fitting*).

Here are some examples of how these principles can be achieved in the classroom.

ad 1 *Context-bound constructs*

* By means of arithmetical play-acting (Van den Brink, 1985), e.g. the scene about the bus journey, children can create a context in the classroom and concentrate on a specific arithmetical model (e.g. repeatedly getting on and off) among all the other subjective notions they may have about the bus.

* Artistic decoration. The children were encouraged to decorate artificial arithmetical symbols (such as bus-arrows) by adding and 'trimming' signs from several contexts.

Arrows in a contextual frame work:

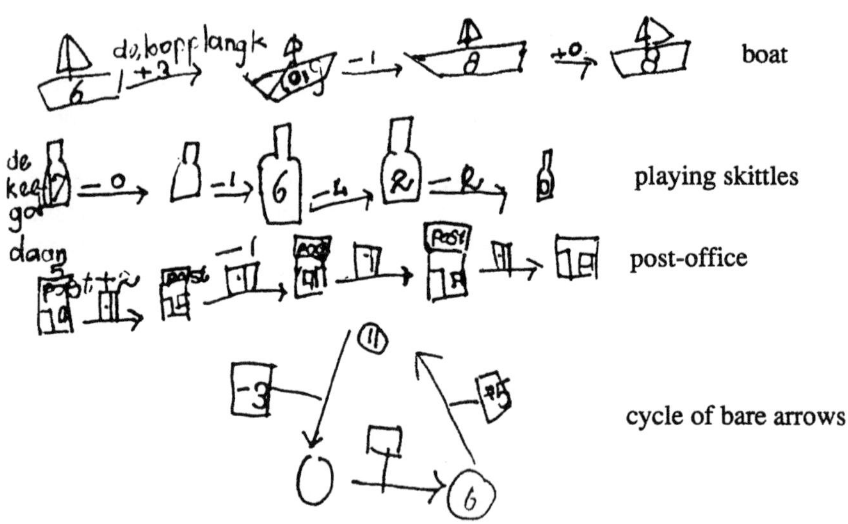

Figure 3.

As a result the children thought up all kinds of contexts to embellish the arrows as if they were dealing with contexts and meanings. Therefore the arrows were *applied* to all kinds of situations (skittles, marbles, stories).

ad 2 *Confrontations*

These took place in a variety of areas:

- o conflicts with the thoughts and opinions of other children,
- o sometimes based on social laws,
- o conflicts with language constructions,
- o conflicts with physical laws.

An example: bathroom-scales.
The pupils have been to the school doctor. They have been measured and weighed. The teacher took her bathroom-scales along. Peter is allowed to stand on the scales. The rest stand round about.

What do you weigh? Peter reads the dial. The teacher asks: 'Can you make yourself heavier?' Peter puffs up his cheeks – no effect. Others make helpful suggestions: 'You must stand on one leg'. Then you do seem to be heavier (is it 'work' or 'weight'?), but Peter cannot make the scales register a different weight. 'Then you must sit or lie on the scales'. 'If you go to sleep, you'll be heavier'. 'You'll have to eat a lot'. All kinds of suggestions and predictions are made in order to try and influence or explain the unchanged weight.

In grade 2 we use two pairs of scales: a child stands straddling the two, one leg on each. Then the weight on each of the scales can be influenced.

A problem that brings new surprises.

In realistic mathematics teaching *conflicting or surprising situations and arithmetical jokes* are used to counterbalance the conventional process of teaching.

ad 3 *Meaningfulness*

There is a difference between 'application of' and 'application in' a constructed reality. We consider the latter to be more meaningful than the former.

An example: Various pocket calculators (Van den Brink, 1986).
We asked each child of grade 5 to bring along a pocket calculator so that we could examine the differences between all these calculators. To the surprise of the pupils the calculators differed not only in color and so on, but also in the results they gave.

Some children got a different answer for the same sum: $4 \cdot 5 - 4 \cdot 5 = 0$ or 80. There would certainly have been problems if the teacher had obtained an answer on his calculator that was different from the pupils. The children's knowledge of mathematics was then applied meaningfully in searching for more sums of this kind and for methods to manage the calculators better, i.e., meaningfully for the pupils.

Not only different calculators but even the same calculator could produce different answers for the same sum at different moments.

When asked to multiply $3 \cdot 3 \cdot 3$, for instance, a pupil was able to conjure up three different answers on his instrument (9, 81 and 27) simply by pressing the keys at different speeds.

The occurrence of different and unexpected results asks the pupil to be aware of his or her intention. Application of knowledge in the constructed reality of the calculator demands meaningfulness.

2.2 Three Arithmetical Topics

Above we gave some guiding principles for classroom teaching. We will now discuss three topics in the Wiskobas project: counting, measuring, adding and subtracting.

We will look at mathematics teaching from the point of view of the 'constructing' pupils by asking ourselves 'how can we create happy, varied and inspiring moments which take the ideas of that pupil into account.'

2.3 Counting

The first kind of counting one thinks of is 'quantity counting'. But in my view it is better to start by considering something completely different. To children, counting – when you come to think of it – is something like:

* calling out words as you walk: one, two, one, two, ...;
* making words: twenty-nine and then? Twenty-ten? '100 and then? One and hundred, two-and-hundred, ... (in Dutch)';
* starting a race: 'Get on your marks – one, two, three!'

Example: the conductor of an orchestra
We let the children beat time (like the conductor of an orchestra) to music on tape. The beat and rhythms varied: one was a folkdance, another Beethoven's Fifth, etc. We observed the strangest of movements. The children divided the music quite differently and counted differently from the way we normally count.
Example: 'hide-and-seek'.
The child who is in turn has to count in tens: 10, 20, 30 up to 100 before seeking.

We call the counting in these examples: *acoustic counting* (Van den Brink, 1984). This kind of counting has very little to do with quantities. Acoustic counting is linked with movement on the one hand and with sound systems on the other. But acoustic counting is not simply a technique for comparing collections of objects.

We have to introduce the counting of quantities as a rather artificial application of acoustic counting. Just note the counting mistakes toddlers make while walking: '1, 2, 3, 4, 5, 6, seven'. Do you take 2 or 3 steps at number 7?

From acoustic counting to quantity counting via conflicts

A didatical problem is of course how to progress from acoustic counting to quantity counting. One of the most important teaching methods for letting children see the differences between these two sorts of counting is to create *conflicts or surprises*. Unlike what they do in acoustic counting, the children now have to take only one step or object forward at each number word and they have to stop counting at the end of a line of objects, and the counting words they used should all be different.

Now we propose to use these properties of counting quantities to create conflicts and surprises, in order to make quantity counting fit into the children's acoustic counting.

Example: A row of 10 counters
The children are given a row of 10 counters each. You do the counting, the children point to the counters. Then you do something silly: after 10 you go on counting or you leave out one of the numbers under 10.

To the dismay of some children. How do you calm down the consternation? Some children point up the missing counters in the air, others count the counters over again, others stop angrily. 'What's the matter? What did I do wrong?' The children will begin by themselves to recognize the essential features of quantity counting: that you have to stop at the end, that you can't leave out a number.

Example: string of beads
The situation can be repeated with a string of beads. But now you and not the children point to the beads. The children have to count out loud. Again you create conflicts by leaving out beads or by going on and on, pointing the beads in a circular way twice or three times. 'How long am I to go on with this?'

Moving away from natural counting

In the long run, natural counting (1, 2, 3, ...) is uneconomical. Large quantities have to be structured, e.g. divided into groups of 5 or 10. Then you can count more easily in tens. Arithmetic problems which are set later on, such as 23 + 92 = are almost impossible to solve by natural counting. In other words, we must equip children with the means. E.g. acoustic counting in tens (10, 20, 30) to move away from natural counting (1, 2, 3, ...).

Example: Moving Collections
Stationary collections are easy to count. But counting moving groups gives problems. For instance, moving children in the playground are divided into groups ('first count the four beside the tree'). Larger units are taken. In addition to acoustic counting in tens, a didactical aid is not to show certain parts or sections.
Examples: Invisible quantities are important.

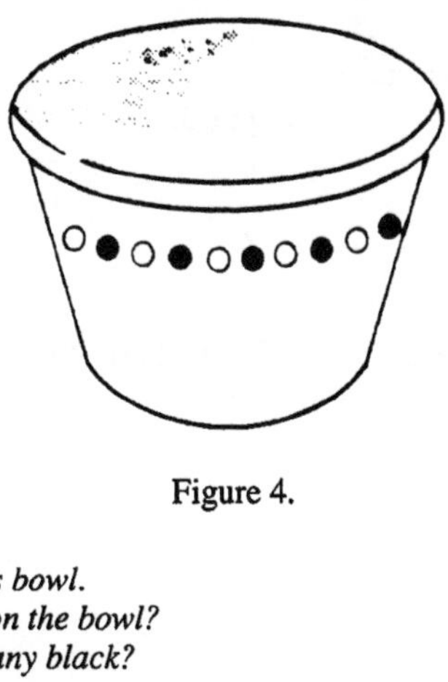

Figure 4.

You can see 10 circles on this bowl.
How many circles are there on the bowl?
How many are white, how many black?

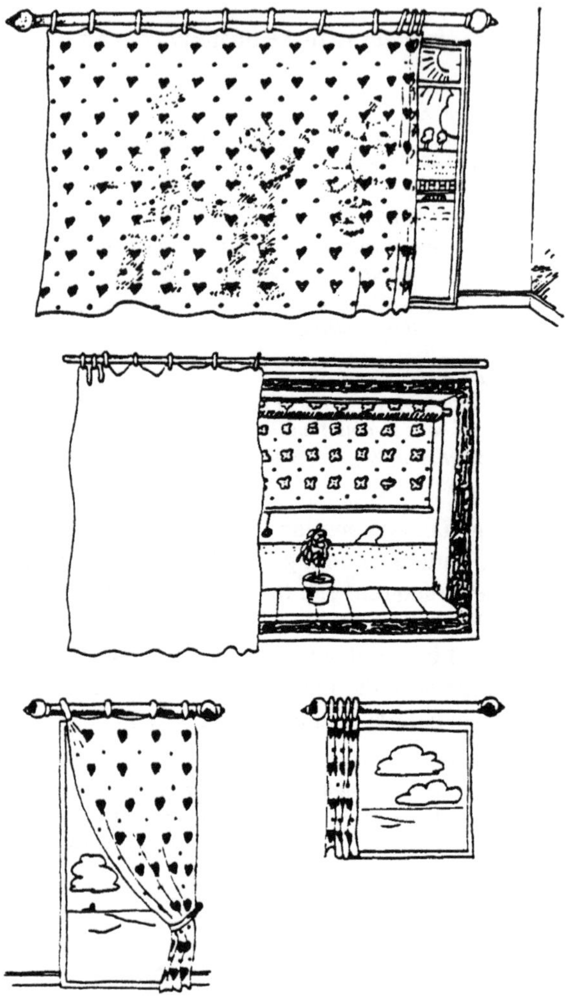

Figure 5.

The flower-patterned curtain.
How many flowers do you see on the curtain?
How many will there be altogether?

Such situations are familiar to children and are within their grasp. They stimulate children to construct a structure.

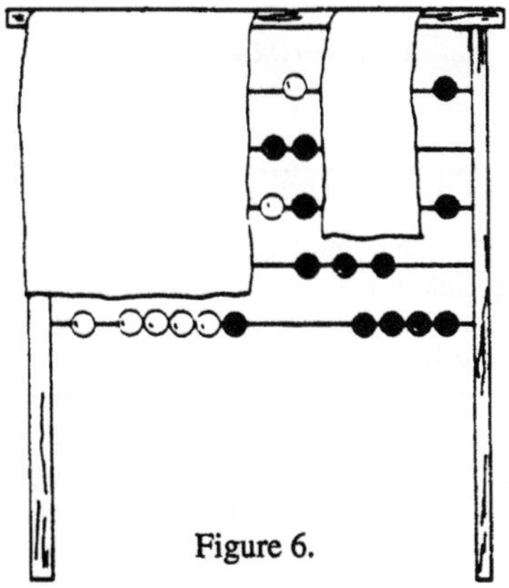

Figure 6.

How many beads are there on each rod of the counting-frame?
How many are white, how many are black?
How many beads cannot be seen?

You see it is not always possible to give the exact numbers - it depends on your imagination.

By 'reading off' the numbers of beads on the counting frame you can count: '10, 20 and 21, 22, 23'.

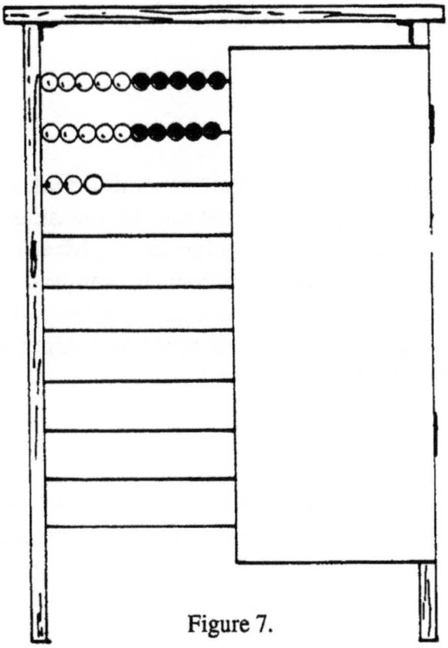

Figure 7.

How many beads will there be behind the panel?

Characteristics of DR

In the rest of this section about counting we shall consider some important characteristics of DR (cf. Treffers, 1987).

* Starting from contexts with which the children are familiar or can be made familiar (cf. Cobb, 1987). For instance, link in with the familiar acoustic counting.
* There are 'means' present at the beginning which can help the children to make progress later. Think of the counting in tens during hide-and-seek and the return to this type of counting during the counting of beads on the counting frame.
* Great importance is attached to things that children produce themselves (Von Glasersfeld, 1988). Think of all kinds of structuring while you're working with invisible quantities.
* Teaching is interactive. By means of class discussions and the performance of counting tasks children can experience the importance of cooperation. Think of counting moving collections.
* Subjects can be interwoven. For instance, acoustic counting can be combined with the teaching of music and movement.
* In my view the most important characteristics of realistic arithmetic teaching are the conflict- and surprise-situations. They are educationally important in that they make the child become aware of the essence of the concept or situation. Sometimes you just have to do silly things in order to induce children to consider counting of quantities.

The properties of a specific mathematical activity, in this case the counting of quantities, have to be used as a basis for a conflict. The properties should not be imposed but must be discovered to be useful and meaningful to the pupils themselves.

2.4 Measuring

When you hear the word 'measuring' you generally think of measuring length. But there are many other magnitudes that can be measured as well as length: surface, content, weight, time, money, chance, ... Measuring also includes a variety of activities: various measurement techniques, choice of unit of measure, giving something a structure to make it measurable, etc. It is not easy for a child to perceive the objective of all these measuring activities. Therefore we must pay attention to this matter.

'Natural' Measure

Over the years a particular method for teaching measurement has evolved. When a magnitude such as length is introduced, the teacher starts from a so-called 'natural measure'. The length of the table is measured not in centimetres but in matchboxes. The teacher then moves gradually towards a standard measure (e.g. a ruler divided into millimetres or a measuring-staff divided into centimetres).

The natural measures however have something *unnatural* about them: can a table be measured in matchboxes? Yes, it can, but it is not the most obvious method to choose. Measuring a table in 'handswould be another 'natural' measure. But hands are not all the same size. Some materials (elastics, etc.) cannot really be used as 'natural measures' because the unit of measure is not fixed. You can't measure length with a piece of elastic. Nevertheless, my mother always 'measured' my waist with a piece of elastic when the elastic broke in my sport shorts.

So what does 'natural' mean? The naturalness of the unit of measure or of measuring activities apparently does not depend on the use of random objects, but depends on the situation or context in which the measurement is made.

The objective of the measurement is supported by the context. As a general rule we can say that it is the measuring context which makes a chosen unit of measure or measuring activity 'natural' for a child.

Measurement Contexts

Measurement contexts give the reason *why* you measure something in a particular way. Such contexts can expand with different activities. There are at least three kinds of activities:

* Performing actual measurements;
* Reasoning about measurements;
* Investigating magnitudes.

Length and distance

* Performing measurements

Example: 'I know what to do', says Dirk, a 5 year-old toddler and he walks along the corridor with a wooden measuring-staff. Here and there he lays it on the ground, but he doesn't keep it straight. He's measuring the length of the corridor in his own way.

Anneke copies him. Neither toddler counts during the measuring.

When I mention, that Dirk has laid down his measuring-staff much more often than Anneke, they're not concerned. They feel they've really 'measured' the corridor. It's quite different when they have to walk to the opposite side of the corridor taking 'great big' steps. They do their utmost to do this in as few steps as possible.

The children really 'feel' the problem: the bigger the steps they take, the fewer are needed. Now at least they're going towards their 'target' in a straight line. The objective of the standard measurement technique is gradually being achieved: in a straight line. Only they still do not count their steps. 'Counting' for them still has nothing to do with 'measuring'.

This is not really surprising, for they don't count for instance when they use a way of measuring to put one foot behind the other in order to know who may start a game.

In grade 1 counting was discussed in connection with measuring. 'The children crossed the classroom, at first they took 12 steps to reach the other side. But when I asked if they could do this another way, some boys crossed in 6 steps and some girls took 35 tiny steps to cross the room. Others stopped counting at 19, because they didn't know what number came next...'.

The conditions governing the performance of measurements emerge through the creation of all kinds of conflict-situations in which it becomes meaningful to measure length in a *straight* line, choose a *fixed* unit of measure and couple measuring with *counting*.

* Reasoning about measurements: measuring by eye. A measuring activity however does not really have to be performed with a measuring-staff or in steps. You can also measure 'by-eye'. This is backed up by children's experience and imagination. But the fundamental idea is again that a measuring context is created.
  Here are some examples taken from experiences in various grades.

Figure 8. Madurodam (grade 1)

How big are you in this picture? (Look at the door.)

Figure 9.

The connection between an object and its surroundings is determined very subjectively (Von Glasersfeld, 1988). This notion is used a great deal in DR.

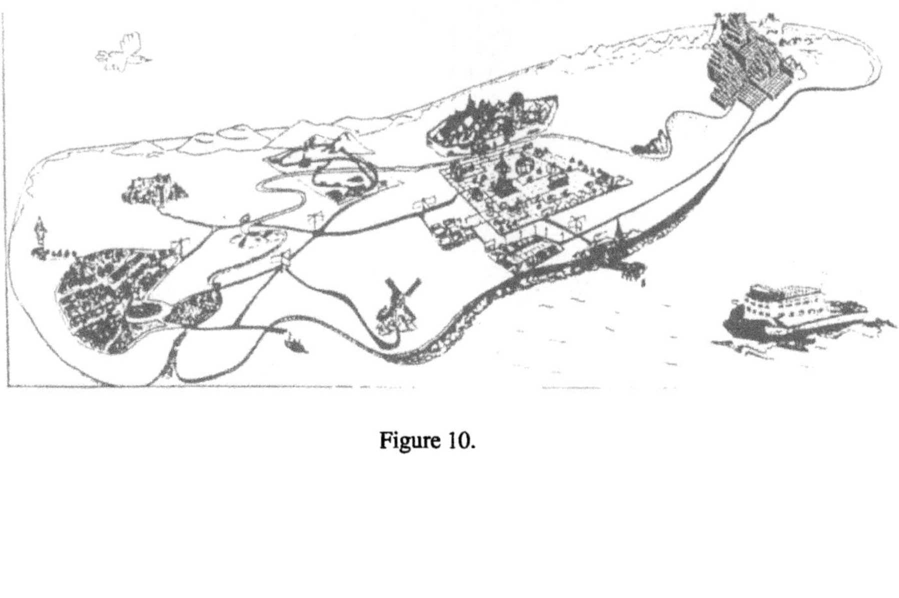

Figure 10.

'How many kilometres do you think the road is on this island?' asks the teacher.
'Four kilometres'.
'That's absolutely correct.'
'And how long is this other road then?' (by comparison 8 kilometres)
'Five kilometres', says the pupil.
'That can't be right, if this road is four kilometres long?'.
'Oh no, then it's about eight.'
'See if you can find a road that's one kilometre long'. ---

The children have absolutely no idea what a kilometre is (who does?). But we can use it in our thoughts: we simply used the four kilometre road as a unit of measure.

Figure 11.

This book contains 100 pages.
'At what page did I stop reading?', the teacher asks.

Figure 12.

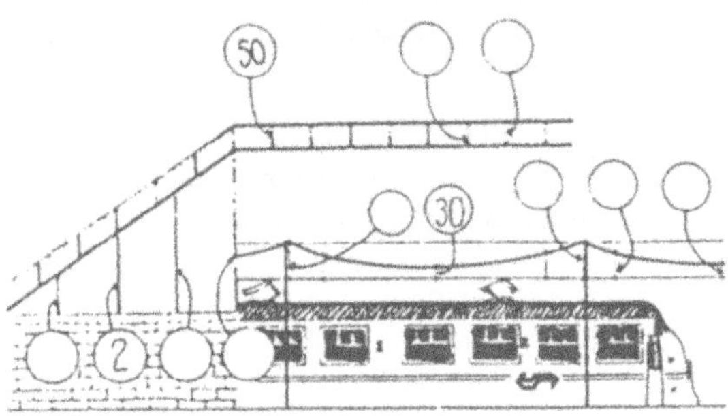

Figure 13.

If we study these examples taken from grade 1 and 2, then it becomes clear that the natural measure is bound to a context or to surroundings and it is the context or surroundings which determine to relations between unit, number and the length to be measured.

* *Investigating the distance: measuring under conditions*
Closely linked with length is the concept of distance; the latter is even more dependent on context.

Example
What is the shortest distance between two points? A straight line?
Is there only one straight line?
In a grid you can certainly draw several 'zig-zag' lines between two points that present the shortest distance.

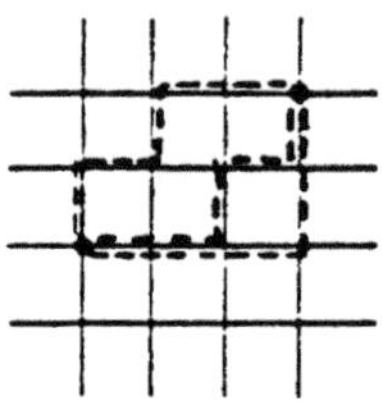

Figure 14.

On a globe of the world you can imagine an infinite number of 'shortest' routes between the North Pole and the South Pole. The line between two points, which has the shortest length is not necessarily the only line (straight, zig-zag or curved).

In towns the shortest route is determined by all kinds of conditions, for instance: all kinds of prohibition notices, speed restrictions, traffic lights at which one has to wait, methods of transport, availability of public transport, etc. In everyday life there are all kinds of examples to show that in a particular situatin the shortest distance does not correspond to the actual shortest length. Walking on a sandy surface is more difficult than on a hard surface and one has to take account of this factor in determining the shortest distance (in time). The older the walker, the longer the distances seem ...

In summary: the concept 'shortest distance' is connected with 'the highest speed', 'in the shortest time', 'in the most convenient way', etc. and is closely bound to the context and to the pupil. We will return to this topic later (section 3).

Other magnitudes

The idea of measuring length as related to measurement contexts, is also applicable to other magnitudes.

* *Surface and Surface Area*

The surface is usually conveyed to pupils directly by means of a grid. Then it is no longer necessary for children to make the surface measurable by constructing a grid themselves.

Example
How many bulbs go into every garden?

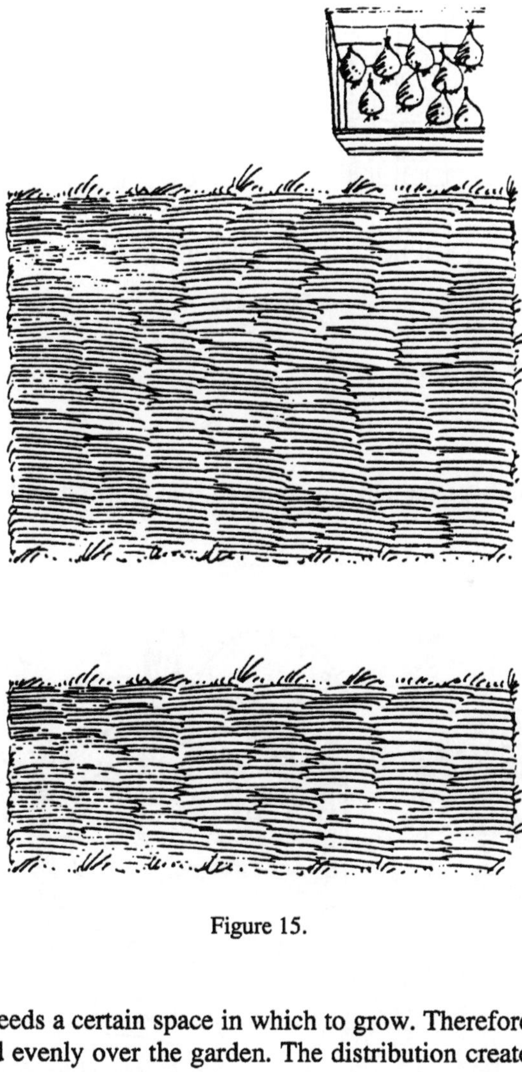

Figure 15.

Every bulb needs a certain space in which to grow. Therefore the bulbs have to be distributed evenly over the garden. The distribution creates a grid, on the basis of which two gardens can be compared. The area around each bulb is, as it were, the natural unit of measure.

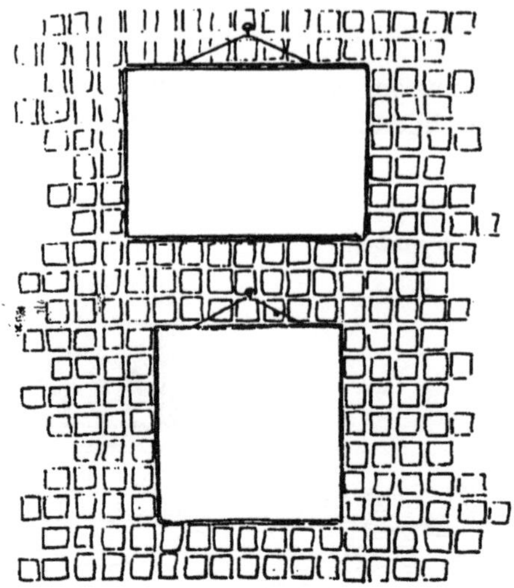

Figure 16.

The lower picture is larger.
Is it really?
* *Time*

If one of the hands of the clock is omitted, the function of the missing hand becomes clearer.

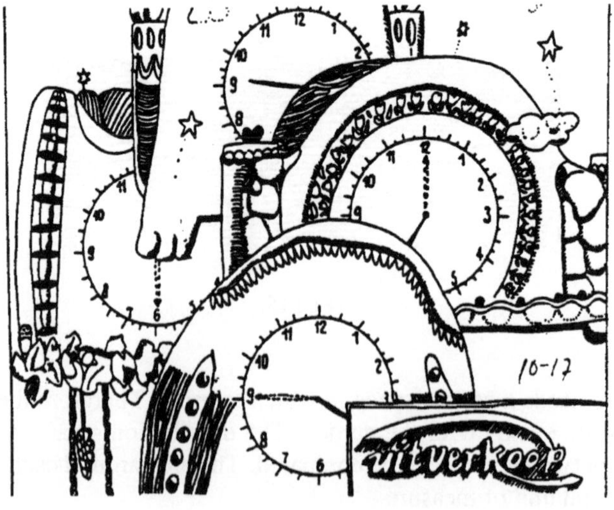

Figure 17.

What's the time on each clock?
In fact the time could be given by the hour hand alone.
A suggestion leading to thought-constructs is the year-clock. One can imagine that the clock-face also gives the months of the year. 1 o'clock: January; 4 o'clock: April. Where will April 15th be?
Show on the year clock when you celebrate your birthday.

* Contents and Weight

With regard to content and weight we recall the example of the closed flask (vase) which was half-filled and the bathroom-scales on which the children had to try and make themselves heavier.

Characteristics of DR

The following characteristics play a role in the teaching of measurement:

* Using *contexts* as a starting point for making things measurable, making units of measure and for measuring activities which are experienced as '*natural*'.
* From these natural measurements the teacher creates *conflict-and surprise-situations* which lead children to form concepts from their *own ideas and experiences.*
* In order to exchange these ideas one needs *interactive* mathematics teaching.
* Apparatus contexts with equipment, such as a clock, bathroom scales, pocket calculators, a half-filled vase, satisfying their own specific laws are often in conflict with the ideas and expectations of children.

2.5 Adding and Subtracting

Illustrating Object Contexts and Artificial Arithmetical Signs

In traditional arithmetic teaching, numbers and signs such as 'equal'-sign, 'larger than', 'plus' and 'minus' are explained with the help of materials: bricks, bars, pictures, etc., most of them being attractively coloured. All of these aids are used for counting or adding and subtracting.

Example
The teacher writes a sum on the blackboard: 4 + 3 = .
The pupils 'lay out' the two numbers in bars (many of the children already know that the answer is 7). They then have to count the total and write down the result in a particular way: 4 + 3 = 7. In this manner the children are first trained to perform the sum with bars and then they are taught how they are to write down the sum. These two activities must be 'matched' accurately.

All kinds of aids (bars, bricks) are used to illustrate the 'bare' written sums. The aim is to bring the children to think in an abstract way and to enable them to apply the sums.

But is it a good starting point to work with illustrative material and artificial arithmetical symbols? The material itself, i.e., the meaning it has for children, plays no role in this approach. Bricks and bars encourage children to build things rather than to add and subtract.

Furthermore, it's not very easy to add or subtract using bricks. Subtracting using bricks is an activity that leads a life of its own, quite separate from adding. Doing arithmetic with bricks means that adding and subtracting are not 'inverse' processes. In traditional arithmetic therefore these two processes are practised separately for a long time. Arithmetic gets 'bogged down' in manipulations which are rooted exclusively in arithmetic. These arithmetical activities can be learned only by long practice with all kinds of sums.

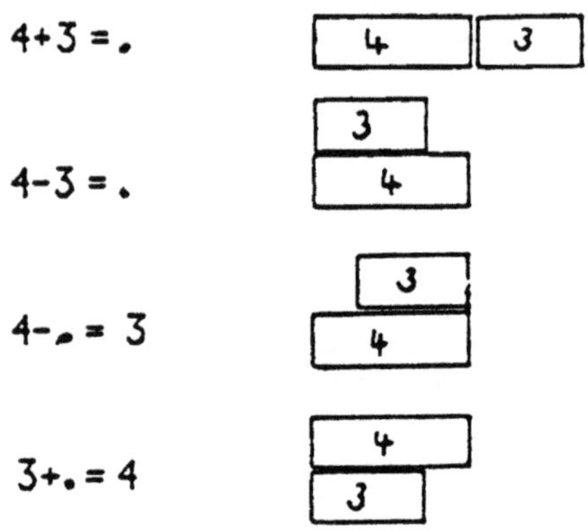

Figure 18.

The several arithmetical notations are accurately matched with several configurations of bars.

Figure 19.

The symbols (+, -, =) are placed unchanged not only between numbers but also between all kinds of objects and graphic representations. The idea is that the 'matching' of numbers to objects helps the children. But in fact a brand rather new silly context is created: doing arithmetic using 'bare' sums is a process that is isolated from the children's world of experience. This is because no attempts are made in the teaching to fit these object-contexts into the children's world of experience, which is much more complex.

Arithmetical Play Acting

Ordinarily children are made to practise adding for seven weeks before being allowed to begin subtracting. Such children are forced to do 2000 sums in the last 3 months of a school-year, whereas they could acquire the same skills in half that time. Nevertheless we should only know that a great deal of practice is needed if children are to go on recognizing sums in certain patterns of bars. But there is a different approach which builds on the playful character of infant teaching.

Infants like acting: they imitate their parents, act the role of children and even of pupils. In class, play-acting can serve as a basis for arithmetic teaching.

In arithmetical scenes (think, for instance, of the bus-trip in the classroom) the material plays an important role of course – but you use it to conceal your objective. The basic idea in such a play is that arithmetic is probably an event in which the child participates (alternately playing an actor and a member of the audience). Sometimes you sit right in the middle of events, sometimes you are more of an onlooker (observer). This alternation is considered to be of great importance for learning arithmetic (Cobb, 1987). Even more important than good material is the image that the material can conjure upon children. The play-acting gives children the possibility to look closely at *one particular activity* (for instance the repeatedly getting in and off the bus). In a play, children never use numbers and processes in a general way, they keep them linked to certain contexts or situations. We can therefore talk about bus numbers, skittle numbers, marble numbers, etc. just as we talk about quantity numbers, ordinal numbers, numerals, name numbers, depending on the situation in which we are using them.

The events that occur in such a situation – continually getting on and off at bus-stops for instance – are recorded in numbers and signs (+ and –) in a notation-with-arrows which remains connected to the bus situation by means of decorative drawings.

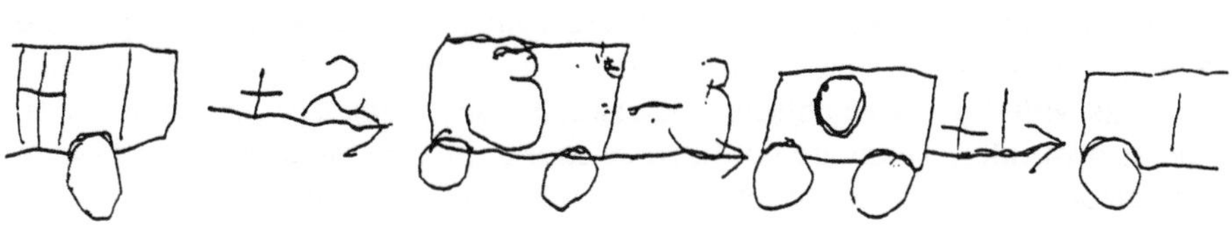

Figure 20.

The arrow notation helps the children *to remember* the course of events during the bus-trip that was acted out in the play and in their eyes this arrow notation is a meaningful arithmetical topic.

What contexts are suitable for arithmetical play-acting?

In our view the following ones are suitable: contexts with people and animals, with games and with equipment.

People and Animals Contexts

There are contexts in which the children themselves take the parts of people or animals. For instance, you tell a story about certain characters such as Peter, John and Joan who are about to travel by bus. The 'events' of the journey are shown on the blackboard by means of bus arrows and numbers. Then the children act the story, referring to the string of arrows. In fact there are many other plays set in human and animal contexts in which children alternatively can take the part of actors and observers. These are based on fairytales (gnomes, giants, fairies), on scenes set in shops or waiting rooms (as patient or dentist) or on tales about animals (pet dogs or cats), etc.

In these situations (social) rules of behaviour often hold and children are familiar with these rules (waiting at the bus-stop, getting on the bus at the front, alighting at the rear, etc.).

By contrast with the illustrative object contexts (such as counting-bars), contexts involving humans and animals are dynamic contexts: the processes which we seek to teach are implied by the context and do not require to be agreed upon.

It turns out that when children are in such a dynamic context they go through the required processes much more quickly than if the context is a static one and the processes have first to be defined. With the bus context children learned to add and subtract in one day, whereas when they worked with counting-bars they practised adding for 7 weeks before they began subtracting.

It is therefore advisable to begin teaching addition and subtraction in a dynamic context involving people and animals.

Toy Contexts

Then there are contexts with toys. These cannot be regarded as people and animals contexts (in which children themselves form the toys), nor are they contexts with illustrative objects. Each situation involving a game (e.g. marbles) has its own rules. These are not fixed arbitrarily in accordance with the laws of arithmetic – which is the case in contexts with illustrative objects – but they are agreed upon by the children.

Toy-contexts are contexts involving marbles, skittles, shuffle-board, dice, the game of goose, etc.

Apparatus Contexts

Contexts with equipment (e.g. bathroom scales, pocket calculator, clocks, the half-filled vase) have their own special rules. The children think these rules exist but do not yet know them. To the children these pieces of equipment are like 'black boxes', the hidden structure of which has still to be discovered. With these apparatus contexts one can stage exciting surprise situations in the classroom and these then constitute the ingredients of the arithmetical play acting.

Decorating Arrow-Notation With A 'Context-Fringe'

Another striking feature of realistic arithmetic teaching is that it is assumed from the outset that artificial arithmetic symbols can be artistically adorned with what we could call the 'context-fringe', in other words with all kinds of 'decorations' which the children invent in connection with a context. These decorations represent the deviations from the arithmetical models which we are trying to present in the context.

For instance, the children let passengers get on *and* off at each stop, whereas – for arithmetical purposes – we let passengers either get on *or* get off. The children used lots of words for addition: get on, more, get in, whereas the only word we thought of was 'plus'.

The children ornamented the arrows for instance with wheels or they added some other extras. For us the plain arrows sufficed. Each of these fairly primitive linguistic decorations, however, gave rise to new educational insights. Adding decorations of the plain arrows seemed to be a very important didactical aid: by means of various decorations the arrows could be linked to innumerable contexts. The children thoroughly enjoyed decorating the arrows.

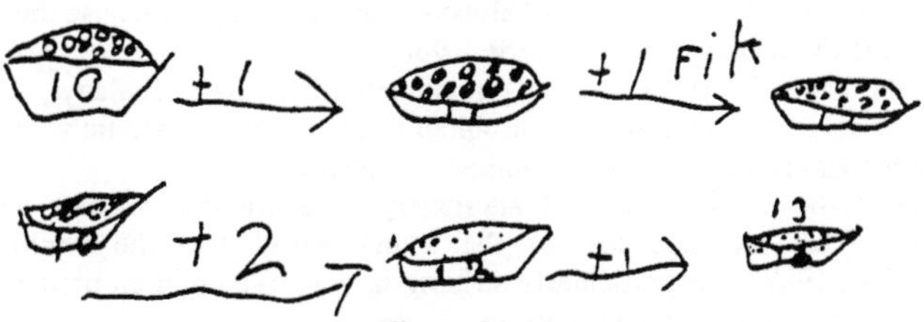

Figure 21.

And in this way the children themselves 'transferred' the arrows from one context to another and abstracted them. Instead of saying 'get on' they sometimes said 'plus', instead of drawing a bus-arrow they finally noted down the problem in plain arrows.

And later as:

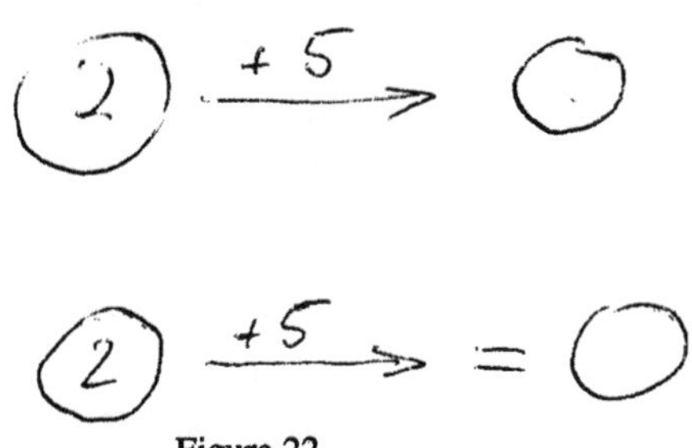

Figure 22.

and finally $2 + 5 = \ .$

Figure 23.

Through the omission of some context-fringes (wheels, bus-stops) and the addition of others (e.g. placing the equals sign after the arrows) the transfer was made from expressive contexts to ordinary arithmetic. The advantage of the meaning-changing trimming on the plain arrows was that the arrow-notation was rendered abstract, but the reverse process (from plain counting to expressive contexts) could still occur.

Later on, whole contexts could be conjured up with one word and served as applications for traditional arithmetic. Words were more than simply carriers of meaning, they carried instructions about choosing from various contexts (Von Glasersfeld, 1988). The official arithmetical language with the equal-sign fails to give pupils this opportunity to introduce their own expressive decorations. Traditional arithmetic is concerned with arithmetical activities with counting and not with the representations that children have of their activities.

In summary, the transfer from one context to another via the arrow-notation is characterized by three features:

1. The original 'bus meaning' of the arrow notation *faded away*. Even in normal speech one is not inclined to conjure up in one's mind a detailed and specific situations for every story or sentence. One doesn't always construct images because this would be too cumbersome (Von Glasersfeld & Steffe, 1986).
2. But on the other hand the symbols remained *transparent* (Polanyi, 1962). By the addition of an arrow to an ordinary problem in traditional arithmetic with an equal sign the former 'bus-meaning' is easily conjured up again.
3. As a result the children worked alternately in expressed contexts and traditional arithmetic. This alternate use of the living world and the world of symbols (Freudenthal, 1987) was particularly striking in what the children produced when they had to think up problems themselves.

Static Contexts: Bases For Conflicts, Surprises and Arithmetical Puzzles

Not all arithmetical problems can be interpreted as events in dynamic contexts. Sometimes they involved descriptions of static situations. When, for instance, I say that 'John has 5 marbles and Peter 8' and then I ask how many more marbles Peter has than John, then it's difficult to speak of an 'event'.

The same holds for problems such as 'There are 11 children in the classroom, 7 of them are girls. How many boys are there?' The numbers here describe *static* situations, not events. If a child is to solve such problems however, it must turn them into an event, otherwise it is not doing arithmetic. If the situation is introduced as a puzzle or a joke, then the child has an extra stimulus to start working in a static situation.

Invisible quantities, situations involving magic tricks and the existence of several possible solutions are important instructional aids to stimulate children's interest.

Examples

Figure 24.

'How many dots?'
'How many dots are on the other side?

Figure 25.

'Does every pussy have a 'doggy' friend?' 'Of course, the numbers are equal as you can see'. After the animals had been cut out, one pupil looked around on the floor: she missed one of the dogs.

Figure 26.

In the school TV for grade 1 two clowns let a donkey come in that could only bray 'five'. The clowns asked the children to help them with this scene called 'The Clever Donkey' by sending in all kinds of sums to which the answer was 5. They intended to put these problems to the donkey during their act and as a result their donkey would appear to be very clever.

Figure 27.

The question 'How many puppeteers are there in the puppet-theatre?' will certainly lead to different one-to-one correspondences and therefore to different answers.

Clearly then we can present static contexts as surprise- or conflict-situations in order to stimulate children to conjure up the situation in their mind.

Characteristics of DR

In summary, the following realistic teaching charcteristics can be found for addition and subtraction.

* People and animals contexts, wherein the pupils play the role of actor and observer, and toy-contexts, in which the game follows its own rules, are to be preferred to contexts involving illustrative objects.
* These 'static contexts' can be helpful in realistic arithmetic teaching if they are presented as conflict- or surprise-situations. They can help children to gain a better grasp of arithmetical concepts.

* The transfer to 'abstract' traditional arithmetic is stimulated by these conflicts and surprises. The decorating of the arrow-notation with various 'context-fringes' means that the arrow-notation is not only attached to a variety of contexts, but it also stimulates the *transfer* of the arrow-structure between contexts. As a result of these switches between expressive and bare contexts through the decoration of the arithmetical symbols, the meaning of the original images becomes blurred, the arithmetical symbols used remain transparent and arithmetical language is used both in the world of symbols and in the living world of children.

III. RADICAL CONSTRUCTIVIST VIEWS AND DIDACTIC REALISM

What contributions can RC make to the teaching of arithmetic and in particular to didactic realism? Every researcher/developer will probably give a different answer to this question. I have found a number of striking ideas described in RC; these have forced me to reflect about the developed realistic teaching programme. I refer to these ideas below.

3.1 Constructs of the World Outside Ourselves

Knowledge of the world outside ourselves refers more to 'the construction of paths of action and thinking that an unfathomable 'reality' leaves open for us to tread' than to what is true or false in that objective reality (Von Glasersfeld, 183, p. 2).

In other words, knowledge of the outside world depends on the way in which and the means by which we have acquired this knowledge (by what senses, for instance?)

Von Glasersfeld and Cobb (1983) point to the disturbance that can be caused by one and the same phenomenon if it leads to different experiences at different moments (e.g. there is a path through a cornfield at one moment and none the next). This disturbs the 'repeatability' and that is sufficient reason to do further investigation with other senses (p. 222). Is knowledge acquired through action or contemplation? Closely linked with the ways of construction is the difference between 'actors and observer's perspectives' (Cobb, 1987).

In these theoretical observations we could see a justification for the arithmetical play acting. In it children play the part of actors and observers alternately, different senses are stimulated and constructs are made with the imagination.

3.2 Views of the Outside World

Something that is typically 'radical' is the constructivist criticism levelled at persons taking part in a discussion in which 'the participants believe that *their* meaning of the words they have used are fixed entities in an objective world outside the speakers' (Von Glasersfeld, 1983, p. 10). Constructivists assume 'that speaker's meanings cannot be anything but subjective constructs' (p. 10) for '(...) there are always differences that I consciously or unconsciously *disregard* (...)' (p. 4). Our idea of the context-fringe is based on this notion. We regard a context as a conglomerate of children's notions which emerge during the play or in conversations in the classroom. These notions consist of models that are important for arithmetic, but they also consist of decorative context-fringes.

We noticed that children sometimes neglected the arithmetical models, preferring the fringes, which represented their own experiences. These fringes formed the subjective surroundings in which they could place the model and they were therefore of great importance to them. But the 'fringe' also seemed to be useful for the arithmetic teaching in that it let the arrow-notation be changed into abstract concepts. By freely decorating the arrows with elements from various contexts the children were in fact dealing in expressive contexts while they were using the bare arrow-sums. Von Glasersfeld stresses this via his notion of the 'subjective environment': object and background are interdependent and neither is less subjective than the other (1983, p. 5). This aspect is used in DR: think for instance of the house in Madurodam which very soon turns out to be a doll's house.

We know the world through our own experiences and assume erroneously that our thoughts about the world are the world itself. This is one idea of RC that we experienced strongly when the children beat time to various types of music. We saw the strangest of movements. Suddenly, I realized that the children were not beating time incorrectly but were 'feeling' a beat that was totally alien to me.

Moreover it is understandable that one's first opinion is considered to be the correct one. Children have a similar experience. Each child thought for instance that the vase was half full and many children thought that you could make yourself heavier on the bathroom scales. But unlike grown-ups, children are inclined to revise their opinions; as a result of this, children in a classroom can construct a kind of consensus about how they as classmates (will) think about the objective outside world. In realistic teaching, mental objects (Freudenthal, 1984) are of intersubjective origin.

3.3 Fitting Knowledge – Meaningful Knowledge

"(...) 'understanding' is always a matter of fit rather than match (...)" (Von Glasersfeld, 1988, p. 10). Knowledge must *fit into* rather than *match* the aims and experiences of children and teachers.

We interpreted this constructivist view as follows: 'see to it that constructs are meaningful (full of sense) for children'. The use of arrow-notation for remembering a bus-trip, the need to have pocket calculators that give the same result for the same problem, the need to compile an arithmetic book for other children. These are all examples that ensured that children's own arithmetical constructs in these teaching situations were experienced as being important and therefore meaningful.

3.4 Conflicts and Surprises

Children must be given the opportunity to modify their notions. But how are they to be stimulated to do this? From the beginning DR has stressed the importance of *confrontations* for learning arithmetic gradually in a course. Conflict- and surprise-situations are developed and arithmetical jokes are compiled. This is also a viewpoint of RC. As well as conflicts with physical and social laws and other views. Cobb (1987) mentions conflicts with abstractions. Von Glasersfeld and Cobb (1983) point out that children have conflicts with their own experience when experience is not the same in *almost* the same circumstances. Think for example of the different results obtained for the sum $3 \times 3 \times 3 =$ when a pupil pressed the keys of a calculator at different speeds (9, 81 or 27).

According to RC a phenomenon can never be repeated in an identical form because one neglects some details in the reality, stresses others (e.g. the context-fringe) and because one cannot claim to know the whole of reality.

This is a philosophical basis for the conflict- and surprise-situations with different results such as developed in DR.

3.5 Language

The linguistic views adopted in RC are particularly attractive as a basis for considerations of this aspect in realistic teaching.

We mention some examples. '(...) A linguistic expression (...) has the potential of being interpreted in terms of actual representations but can also be stored without the interpretation (...)' (Von Glasersfeld & Steffe, 1986, 213). In other words: when language is used, often no detailed representation of sentences is evoked, although the individual has the ability easily to evoke such a representation.

Scenes, formerly acted out in arithmetical play acts are referred to by a single word. But even 5 months later pupils could still describe them in detail. The meanings of the words and symbols gradually grow fainter, but nevertheless the meanings can still be evoked.

This situation is also described by Shannon (1948) and can be summarized as follows: the physical *signals* that pass between communicators do not carry what is ordinarily considered 'meaning'. Instead, they carry *instructions to select* particular meanings. Polanyi means that the symbol even serves as a substitute for a meaning. He refers to this as the 'transparency of symbols'. 'The most pregnant carriers of meaning are of course words of a language, and it is interesting to recall that when we use words in speech or writing we are aware of them only in a subsidiary manner' (Polanyi, 1962, p. 57). The pupil can let the meaning shimmer through the symbols.

Our experiences with the referential arrow-notation fit into this view perfectly. We found, for example, that the special meaning of a word like 'get on' in the bus context grew fainter. But we also discovered that it was used as 'plus' but actually said as 'get on'. Freudenthal speaks of 'anontology' (1973): the links with the original situation have to be severed certainly, not artificially but more in the linguistic manner intended by RC, as outlined above.

3.6 Final Remarks

With these examples we have tried to indicate how RC has stimulated us to think about arithmetical topics in a realistic way. On the other hand the realistic teaching ideas have taken root among the radical constructivists (Steffe, 1988).

We hope that this chapter will make people think more deeply and carefully about the consequences for practical mathematics teaching in the classroom.

NOTES

1. Having a birthday. We were about to go and see a friend who lived on a farm. My son had just celebrated his birthday at school when his three-year-old sister asked him: 'Are you also going to have a birthday on the farm?'. To her 'having a birthday' was apparently dependent on place: supplying children with cake(s) at school, the party at home. The invariance in time of having a birthday is not an important factor for children.
2. The term 'to realize' has other meanings besides 'the making of cognitive representations'. A second meaning is the actual output of for instance a plan: 'The plan is being carried out'. This meaning is important because you can tell from the plan that has been carried out that the original idea was in fact 'realisable'. A third meaning of 'to realize' is: to make something your own (so that it belongs to you). According to this meaning an idea can be regarded as a personal possession if it can be used or applied independently. A fourth meaning is 'self-realisation' (Dewey), 'self-activation' (Maslov). These are terms that refer to the development of individual qualities.
3. For a three-dimensional approach to mathematics teaching in the Wiskobas-project, see the book by A. Treffers, 1987.

REFERENCES

Brink, J. van den: "Acoustic Counting and Quantity Counting", *For the Learning of Mathematics* 4 (1984), 2–13.

— "Numbers in Contextual Frameworks", *Educational Studies in Mathematics* 15 (1984), 239–257.

— "Staging Arithmetic: A Suggestion for the Start of Mathematics Instruction," *For the Learning of Mathematics* 5 (1985), 35–37.

— "Calculators and Realistic Arithmetic Instruction Developing the Calculating Automation", *Proceedings of the Tenth International Conference of the PME*, London, 1986, 475–480.

— "Children as Arithmetic Book Authors", *For the Learning of Mathematics* 7 (1987), 44–47.

Cobb, P.: "Information-Processing Psychology and Mathematics Education, A Constructive

Perspective," *Journal of Mathematical Behavior* 6 (1987), 3–40.
Freudenthal, H.: *Mathematics as an Educational Task*, Dordrecht, 1973.
— *Didactische Fenomenologie van Wiskundige Structuren*, Utrecht 1984.
— "Theorievorming bij het Wiskundeonderwijs. Geraamte en Gereedschap", *Panamapost* 5 (1987), 4–15.
Glasersfeld, E. von & P. Cobb: "Knowledge as Environmental Fit," *Man-Environment* 13 (1983), 216–24.
Glasersfeld, E. von & L.P. Steffe: "Composite Units and the Operations that constitute them", *Proceedings of the Tenth International Conference for the PME*, London, 1986, 212–216.
Glasersfeld, E. von: "Environment and Communication", *Proceedings of* the Sixth ICME, Budapest, 1988, 1–13.
Polanyi, M.: *Personal Knowledge*, Chicago, 1962.
Shannon, C.E.: "The Mathematical Theory of Communication", *Bell Systems Technical Journal* 27 (1948), 379–423, 623–656.
Treffers, A.: *Three Dimensions. A Model of Goal and Theory Description in Mathematics Instruction – The Wiskobas Project*, Dordrecht 1987.

TWO LAYERS OF CONSTRUCTIVIST CURRICULAR INTERACTION

ROBERT G. UNDERHILL

Virginia Polytechnic Institute and State University, Blacksburg, VA

In this chapter I will try to help you construct a "feel" for constructivist teaching and learning by examining two mathematics research and development projects first explored at the London meeting of the International Group for the Psychology of Mathematics Education (Underhill, 1986). I will also draw upon the reflection papers written by classroom teachers prepared while enrolled in my undergraduate and graduate classes. Their papers will at once communicate to you the types of activities which help them begin to grasp and appreciate the nuances of meaning constructed by them and by some of their students. Such expositions will also help you gain some sense of the powerful insights achieved by classroom teachers as they struggle to make their own meaning from interactions with students and other adults. One very interesting note is that neither of the projects has concerned itself with constructivism. Epistemological questions are often addressed implicitly in curriculum development projects. Rarely are they addressed explicitly. It is therefore of some interest to examine these projects retrospectively. Probably other epistemological positions could be used to discuss the projects, but that is not the point. At issue here is an exploration of the constructivist's perspective. The world can be interpreted through this lens and, I hope, this exposition will help make explicit some of the advantages of doing so.

At the heart of this matter is the need to discuss two layers of curricular interactions: those which bring together teacher educators and teachers, and those which bring together teachers and pupils. It is my intention to make clear that success in constructing mathematical knowledge is greatly enhanced if these two layers of interaction are in epistemological harmony. That is to say, teacher education which is constructivist in its orientation is most likely to produce teacher-pupil interactions which are constructivist in orientation. Another way of pointing to this need is to say that didactic or transmission models used to educate teachers about constructivist learning and teaching are incongruous!

Assumptions

We will proceed under certain assumptions about constructivist learning:

E. von Glasersfeld (ed.), Radical Constructivism in Mathematics Education, 229–248.

1. Cognitive conflict and curiosity are the two major mechanisms which motivate learners to learn.
2. Peer interaction is a major factor in producing cognitive conflict;
3. Cognitive conflict induces reflective (metacognitive) activity;
4. Reflection is the main factor which stimulates cognitive restructing;
5. Items 1, 2, 3, and 4 are cyclical.
6. The cycle always occurs within and is informed by the learner's experience; and
7. This cycle empowers learners, i.e., puts them in control of their own learning.

The set of assumptions can be diagramed as in Figure 1.

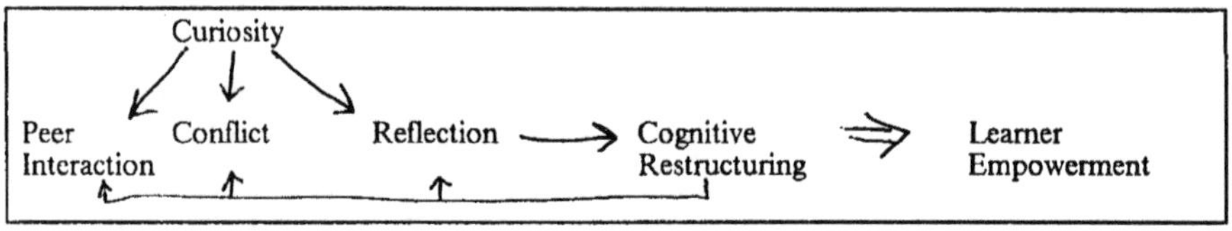

Figure 1. Empowerment which results from learner meaning-making from experience.

Personal experience encompasses the learner's personal history and the activities of the present. Present activities might include text book reading or problem solving, small group work, a lecture, or any of a number of other intended (curricular) or unintended learning events.

Here's how we will proceed:

A. Conflict Teaching: After an overview of this approach to remedial (and original) instruction, we will examine the curricular interactions as they relate to teacher educators, teachers, and pupils. Then we will examine some examples and teacher testimonials related to the curriculum project and from some addition experiences designed to promote cognitive conflict.

B. Algebra Learning Project: After an overview of the key ideas of this project from a constructivist perspective, we will once again examine curricular interactions, examples, and testimonials.

I. CONFLICT TEACHING

At the University of Nottingham Shell Centre in the United Kingdom, Alan Bell and his cohorts have developed an approach to remedial instruction with 12, 13 and 14-year-olds which they have labeled Conflict Teaching. While Bell and his colleagues do not specifically describe their epistemological biases, they cite Piaget in their discussions of cognitive conflict, and their work certainly can be viewed with advantage from a constructivist perspective.

1.1 Description

Bell and his colleagues have sought to develop instructional practices which enhance long-term retention and transfer. They report that following only eight hours of instruction, Conflict Teaching yields better long term achievement than direct instruction. Their approach has been to identify pupils' misconceptions and then to help them overcome the misconceptions by one of two methods. In *direct instruction*, "intensive teaching, focused sharply on known misconceptions was successful in improving performance, but there was very little transfer to points not so strongly focused upon, even though they were dependent on the same general concepts ..." (Bell, 1983, p. 207). Bell reported that when specific misconceptions were the focus of specific instruction that there were specific changes followed by some regression. When "correct procedures were immediately taught for avoiding the errors," the results were not as impressive as when conflict teaching was used.

Conflict Teaching "was intended to involve the pupils in discussion and reflection of their own misconceptions and errors, thus creating an awareness that new, modified concepts and methods were needed. There was, therefore, a 'destructive' phase, in which old ideas were shown to be insufficient and inaccurate before new concepts and methods were introduced" (Swan, 1983, p. 211). On the other hand, "the *positive only* approach made no attempt to examine errors, and in fact avoided them wherever possible by teaching the pupils to use simple and efficient methods from the start" (Swan, 1983, p. 211).

The sequence of the conflict lesson has been described two different ways. In his 1983 paper, Swan described four phases:

1. Intuitive – perform tasks to expose misconceptions;
2. Conflict – perform same tasks, having been provided by the teacher with an easy, understandable method;
3. Resolution – a discussion of discrepancies between phases 1 & 2;
4. Reinforcement – practice of the phase 2 method.

In a Teachers' Handbook (Bell, 1983), this procedure was described in three phases:

1. Open Activity – designed to elicit misconceptions; best to use groups or pairs;
2. Discussion – class discussion of correct and incorrect procedures; probing occurs;
3. Consolidation – a few practice exercises from various perspectives; furthers reflection and understanding.

Research results have indicated that "the amount of cognitive conflict generated in discussion is significantly related to pupil gains on tests" (Bell, 1984, p. 55). Two methods found to be especially useful in provoking this conflict are (1) having students make up problems, and (2) having students mark real or fictitious homework (Bell, 1986).

From a constructivist perspective, it seems fairly straightforward as to why conflict teaching produces better long term results than direct instruction. When pupils develop misconceptions, they often hold them for a long period of time. The misconceptions are part of their realities. The pupils have constructed their own meanings. When teachers use direct instruction, they attempt to give pupils correct concepts (from the teacher's perspective) to replace pupils' "correct" concepts, i.e., pupils' perspectives which they believe to be true but which the teacher believes to be mathematically incorrect. The pupils may remember the teachers' methods long enough to pass tests, but because the teachers' ways have not been understood and integrated into the pupils' existing cognitive structures, the misconceptions (from the teachers' perspectives) win out and endure in the long term. On the other hand, the cognitive dissonance produced by the conflict method causes pupils to reflect and reconceptualize, thereby creating new, correct conceptions to replace now-understood-to-be-faulty ones. This produces higher stability for the new, correct conceptions.

II. CURRICULUM INTERACTIONS

The objective of this section is to demonstrate for the reader one way to present Conflict Teaching in a teacher education program in such a way that it will dually impact the activities of teacher education majors and of the activities of the classrooms in which future teachers will teach, i.e., to impact the teacher as a learner so that the teacher can *experience* Conflict Teaching and to impact the teacher as a professional so that the teacher can *use* Conflict Teaching as a pedagogical tool. If we truly believe in a constructivist perspective, the presentation should impact *both* levels by operationalizing the peer interaction/conflict/restructuring cycle presented early in this chapter.

2.1 Teacher/Pupil Interaction

Conflict Teaching presents old (or new) mathematical content to learners in a way which is designed to induce conflict. In both descriptions of the strategy, the learners start with their personal histories, and new experiential data are provided (problems already worked with misconceptions) or created (learners solve some problems). Then, there ensues peer interaction designed to induce cognitive conflict by discussing why problems are right and wrong. This discussion is usually continued at the whole class level after the pupil pairs or small groups complete their discussions. During these two levels of peer and teacher interaction, the cognitive conflict which is produced within learners leads to reflection/metacognitive activity and results in cognitive restructuring. This process can be described as altering existing cognitive structures, or altering beliefs about mathematical relationships, or eliminating mathematical misconceptions. In any case, the pupils become *empowered* because the teacher has structured the learning environment in such a way that pupils learn new mathematical relationships, or change their own cognitive structures, or change their beliefs, or eliminate their own misconceptions through their own metacognitive activity. Pupils have been placed in charge of their own learning; this is empowering!

2.2 *Teacher Educator/Teacher Interactions*

If teacher educators are *practicing* (rather than theoretical) constructivists, then they ought to exemplify in teacher education practice what they believe about human learning at all levels. Thus, the cognitive restructuring cycle should be incorporated into the teachers' learning about conflict teaching.

In practice, there are two different sets of knowledge or beliefs at the teacher education level: (1) knowledge or beliefs about "how to," and (2) knowledge or beliefs about "whether to." As the competency-based teacher education movement aptly demonstrated, "how to" without "ought to or best to" will yield "teachers who can" but carries with it no guarantees that "teachers will." Thus, the teacher educator must attend to both the *ability* and the *will* to do.

The means I have found effective are as follows:

Week 1:	Deliver a mini-lecture about Conflict Teaching; provide an article to read and a reflection paper prepared by one or two teachers from a previous class; ask each teacher to choose a math topic and develop either a list of misconceptions or problems which will elicit them; give them sample lessons prepared at the Shell Centre;
Week 2:	Review the mini-lecture; have the teachers discuss their misconceptions or sample problems in groups of three and then have a whole-class sharing of ideas; provide a couple more articles; ask them to try their topic or one from the Shell Centre lessons during the next week and write a 1–2 page reaction/reflection paper;
Week 3:	Use groups of three to share reactions; then have a whole-class discussion/sharing; collect the reaction papers to look at during the week;
Week 4:	Repeat most of the activities of weeks 2 and 3 if reaction/reflection papers indicate this is desirable or necessary.

(Note: I do not spend four weeks on Conflict Teaching in the conventional sense. Since I espouse constructivism, I spread all major concepts and procedures over a 3 or 4 week period when classes meet once a week. We work on 3–5 concepts at each class.)

If we examine these activities, we can see that the "how to" is carried out in mini-lectures, reading, actual practice at least once, reading reactions from one or more previous teachers, hearing the ideas and implementation experiences of two other teachers, seeing samples from the Shell Centre, and participating in whole class discussions. The "ought to" or "whether" is handled by demonstrating the effectiveness from Shell Centre research, hearing of the successes and positive reactions from previous and current teachers, experiencing success (this is the reason for the repeat sometimes), and sharing personal successes.

The effectiveness of the cognitive restructing cycle for *pupils* is *witnessed* by the teachers during the implementation of the cognitive conflict model. And the cognitive restructuring cycle is *experienced* by the teachers as they discuss ideas and implementation reactions with colleague teachers.

Examples and Testimonials. The following are excerpts from written work submitted by in-service teachers in graduate courses taught by me in 1987. They were written during course sequences just described. The teachers have not always precisely followed the sequence, but their reactions and activities are quite illustrative.

Experiences of Ruth Siko, Junior High Math Teacher

The day after the class of ninth graders had worked in groups on a set of five inequality word problems, I shared with the class the results of their collaborations. The groups came up with the same answer for two of the five problems, but with the other three problems, I was able to provide some conflict. I was glad for the disagreement because the students generally had been reticent in their groups the day before.

For each of the three problems, I presented the different answers on the overhead projector to the class as a whole. I did not tell them the correct answer but had them discuss each alternative. Then the students voted on which answer they thought was the correct one. In each case, we did not reach total agreement after the first vote. It was at this point that students really became energized as they tried to persuade their classmates. After discussing the last problem, I assigned three more problems which were similar to the first set. Most students were able to do these successfully and with greater ease than before the conflict-discussion.

Basically I felt the students did a good job in discussing the problems and felt that they helped each other select the correct procedure. The students were much more comfortable going over the problems again as a class rather than in their original groups. I think they felt "safer" in the large group with me as a mediator. Not that I told them who was right and who was wrong; I let them decide on the correctness of a solution. As the mediator, I just kept the discussion on the topic of mathematics and cut off the insults. "Gosh, but you're dumb" is a typical remark from these adolescents.

For my second attempt later this week, I'm going to try pairs of students instead of the groups of three and four people, and see if the students feel more at ease.

Experiences of Jo Downing, Middle School Math & Science Teacher

My eighth grade math students have been working on problems involving percents. They exhibited three common mistakes. They are often unable to:

1. change percents to decimals, especially percents which are represented with a single digit,
2. recognize when to multiply and when to divide, and
3. change decimals to percents when necessary.

I gave them several of the following types of problems to work:

What is 9% of 72?
25% of what number is 65?
What percent of 260 is 65?

They worked these problems for approximately 20 minutes. I divided them into groups of four to discuss their answers and to find the way to solve the problems and the way to get the correct answers. I stressed that the answer was not the goal. Rather, the goal was to develop a correct procedure. I also provided questions such as: "How did you work the problem?" and "Why did you do it that way?" to encourage them to talk about problem solving. At first, the students were baffled by this new approach, so as I went from group to group, I demonstrated by asking a few questions to "get the ball rolling" and left the rest to them. One or two students were still set on just getting the correct answer and found it difficult to problem solve.

On their homework assignment that night, all of the students received passing grades where they had failed previously. On the end-of-the-unit test, each student, with the exception of the few bent on correct answers only, demonstrated the ability to solve the problems by using the appropriate procedure.

This was a very rewarding experience for both the students and for me. We were all frustrated in different ways with the lack of success in solving percentage problems. The Conflict Teaching process opened up the communication doors again, gave the students opportunities to explore and solve their own problems of comprehension in a non-threatening situation and deepened their understanding of the math process.

Experiences of Ramsey Teed, Intermediate School Math Teacher

The students were divided into groups of about four each. They were given a grocery store worksheet where they had to buy an item and apply the appropriate sales tax. Each group had to come up with a total bill. They had to give the appropriate amount of sales tax and also state the total price of each item. All of the tax problems had a tax that was a whole percent. They had about 10 minutes to come up with an agreed upon answer. We came together as a group to discuss the various answers given by the spokesperson for each group. The students ironed out most problems. They were then given a problem using 4 1/2% sales tax. The groups were also challenged with a problem that was on sale at $5.00 off.

The students quickly pointed out correct money values to each other. They were able to help each other select appropriate money values and ensured that each person changed his or her percents into appropriate decimal numbers. They gained new perspectives about money and demonstrated reasoning skills.

Experiences of Carol Shields, Middle School Math & Science Teacher

In groups my students pointed out each others' errors with a great deal of enjoyment. Although they tried to use me as a referee, I made them use a diagram instead. They eventually established 4 boxes of candy at 50 cents each were not 54 cents but two dollars.

Another thing I noticed was that it gave my lower students who don't usually achieve a chance to be the "expert". I was worried that only the better students would benefit and the slower students would let someone else do the thinking. However, my other class was at the same lesson and was being taught by the book method only. When they all took their test, my class did better. This could be a result of a lot of different things but the result is worth investigating.

I found that teaching this way fit in nicely with my classroom. When given an alternative way of finding an answer they were eager to show each others' misconceptions. This is not always bad. I think they would rather be challenged by a peer than from an adult. It was a nice combination of whole group lesson with the ultimate in individual help.

Experiences of Sandra Frazier, Gifted & Talented Teacher

I found the Conflict Teaching method unique in its ability to reveal methods of problem solving prior to teaching a method for finding percents of a number. Using adults in the teaching situation allowed me to study the problems of people with experience finding percents of numbers. Through this activity I learned that knowing the procedure for finding percents is not useful if the individual does not know how to change the problem into the form needed for the application of the procedure. I am speaking specifically of my subjects' lack of knowledge of the relationship of mixed fraction percents to whole number percents.

I do not think they thought of 32% as being 32/100, but merely as .32. So when they came to 66 2/3% they did not see it as (66 2/3)/100. I think this is a common misconception. I also thought it was interesting that neither of them attempted to figure the percents using proportions instead of equations.

I thought the variety of working might pose some confusion for them as to what number was the percent of the other. This did not cause a problem and they applied the correct procedure.

I thought Conflict Teaching was a good method to use in teaching percents because it allowed me to understand the thinking they were using to solve the problems. It also gave me an opportunity to see that even when they were unsure of how to apply the standard procedure for solving they tried to use logic to deduce the answer.

This method of teaching allows the teacher to help students see how the problems they understand relate to the ones they do not understand. It seems like an excellent way to capitalize on the students' strengths to deal with their weaknesses. Through working in groups students can see how others solve problems and sometimes come up with the same answer.

I also think clarifying mistakes as they are happening is better for the students because they can avoid repeating mistakes and can correct them while the discussion is ensuing. I think Conflict Teaching is a dynamic learning method that is purposeful in its intentions.

III. CONFLICT TEACHING

I've been talking about Conflict Teaching. Now I'd like to say a few words about conflict teaching (no caps!). Cognitive conflict is the major theme of the constructivist learning dynamic. I've found several other ways to produce cognitive conflict. These strategies are for those who move away from present orientations in increments. These strategies permit you to see the interpersonal dynamics without totally revamping your teaching style. If you are like many of us, you will find these strategies so enjoyable and rewarding that you will use them regularly and will develop other strategies of your own.

3.1 Low Stress Test Returns

This is a technique I've used for a long time. When I give a test, the next day I have the students work in groups of three or four with a blank copy. By the time they discuss their conceptions and misconceptions, I can return their graded tests with hardly any defensiveness of correct answers at all!

3.2 Practice After Mini-Lectures

Many math teachers follow the routine of (1) Review homework, (2) New material, and (3) Practice. An excellent point at which to identify misconceptions is between the new material and practice phases of the lesson. You can simply have the students work two or three problems in small groups or have them work the problems independently and then discuss their strategies in a small group. These small group sessions are often MUCH more effective in identifying and clarifying misconceptions about learning than other strategies I've tried.

3.3 Paired Problem Solving

This is a method of solving problems together which has defined roles for the two people. I first learned of this strategy in *Problem Solving and Comprehension* (Whimbey & Lochhead, 1986). One person is the Doer; this person solves the problem and thinks out loud. The other person is the Listener; this person tries to follow the Doer's strategy, keeps the doer talking, and says "I disagree" if this is true at some point. The Listener does NOT solve the problem independently, and the Listener does NOT explain. Once the roles are learned, this method is a good "window" on metacognitive activity. My students enjoy it and are constantly amazed that others do not think the same as they!

IV. THE ALGEBRA LEARNING PROJECT

At the University of Hawaii, members of the Curriculum Research & Development Group are developing a new algebra curriculum; Sid Rachlin is the project director. The curriculum has several features which seem to fit nicely within a constructivist framework. Eight key ideas will be discussed in this section: reversibility, flexibility, generalizability, time, intuition, problem solving, reflection, and teacher role.

4.1 Russian Contributions

Krutetskii (1976) describes three basic processes of reversibility, flexibility and generalizability in his model of mathematical abilities. These processes are integral components of the Algebra Learning Project as it seeks to develop process-oriented curriculum materials (Rachlin, Matsumoto and Wada, 1985).

4.2 Reversibility

One way to describe reversibility is to relate it to parts and wholes. In traditional algebra instruction, the focus of teaching and learning is unidirectional on combining parts to get wholes and decomposing the whole into parts. Reversibility stresses the BIDIRECTIONALITY of joining and decomposing both conceptual and procedural knowledge.

TRADITIONAL	REVERSIBILITY
1. $-3 \cdot -4 = ?$	1. $-3 \cdot ? = +12$
	$? \cdot -4 = +12$
	$? \cdot ? = +12$
2. $(2a - 6)(3a + 5) = ?$	$(2a - 6) \cdot ? = 6a^2 - 8a - 30$
	$? \cdot (3a + 5) = 6a^2 - 8a - 30$
	$? \cdot ? = 6a^2 - 8a - 30$

This use of reversibility is quite compatible with the Piagetian construct since pupils are moving back and forth between the parts and the wholes. Composition into wholes is paralleled with decomposition into parts. In this work, pupils gain a better feel for the bidirectionality of mathematical processes, sometimes referred to as "doing" and "undoing." The teacher's perspective on reversibility might be the establishment of part-whole relationships. The student's perspective might be a realization that mathematical objects which can be combined can be separated. Krutetskii (1969) says, "By reversible (two-way) associations (and series of associations) we mean those associations in which the thought or realization of the second element (or of the last element) evokes the thought or realization of the first element" (p. 51).

4.3 Flexibility

According to Rachlin et al (1985), "Flexibility was identified by Krutetskii (1976) as the ability to switch from one level of thinking of a problem to another. Kilpatrick (1978) noted that there were two aspects of flexibility in Krutetskii's research: flexibility that can be shown either within or across problems. Within problem flexibility refers to the ease with which a student switches from one method of solving a problem to another method of solving the same problem .... Across problem flexibility refers to the degree to which a successful solution process on a previous problem 'fixes' a student's approach to a subsequent problem" (p. 5).

Within flexibility is exemplified by ease of pupils' use of guess-and-test, computers, traditional algorithms, manipulatives, and drawing pictures and diagrams as alternative methods to solve problems. *Across* flexibility is exemplified by the use of second and third approaches on problems when the approach used on a previous problem is not leading to success. Flexibility of both types is characterized by the use of exploratory moves in the identification of successful strategies.

From the teacher's perspective, flexibility is developed by modeling it in mathematics instruction and by providing pupils with a rich repertoire of problem solving strategies and heuristics. From the pupils' perspectives, they are flexible when they have a variety of problem solving strategies from which they can draw, and can move on to other strategies when the ones they are applying are not leading to success.

From a constructivist perspective, flexibility can be viewed as the ability to create connections between problems and strategies, to see in new problem solving situations structures previously encountered. The pupil can explore relations between previously encountered situations and new ones, to find familiar within the unfamiliar, to apply known routines or subroutines in non-routine settings.

4.4 Generalizability

As with flexibility, generalizability exists in two senses. First, there is the use in the conventional usage of finding what particular instances have in common. The second sense is one which is only lightly touched upon in regular curricula. It has at its core the development of richness and connections among bits and pieces of mathematical knowledge which are often taught and learned with few connections. The conventional curriculum generally does not emphasize this. For example, after teaching acute, obtuse and right angles, most curricula focus on recognizing and labeling these angles. These experiences develop generalized or conceptual knowledge of the first type. To get at the second type, the teacher might ask the following questions.

1. Can a triangle have exactly three acute angles? Two? One? None? Show me examples.
2. Can a triangle have exactly three obtuse angles? Two? One? Show me examples.
3. Can a triangle have exactly three right angles? Two? One? Show me examples.
4. Can a quadrilateral have exactly four acute angles? Three? Two? One? Show me examples.
5. Can a quadrilateral have exactly four obtuse angles? Three? Two? One? Show me examples.
6. Can a quadrilateral have exactly four right angles? Three? Two? One? Show me examples.
7. What combinations of right, acute and obtuse angles can you have in a triangle?
8. What combinations of right, acute and obtuse angles can you have in a quadrilateral?

From the teacher's perspective, the second type of generalizability produces in students a much richer, more interrelated domain of mathematical knowledge by getting students to relate various concepts and generalizations. From the pupil's perspective, generalizability is exemplified by relational learning (Skemp, 1979) and semantic knowledge (Resnick, 1981). From a constructivist perspective, the learner possesses mathematical knowledge which forms a better "fit" (von Glasersfeld, 1983) with the teacher's and the mathematician's.

4.5 Intuition and Time

In traditional curricula, there is a common two day sequence of (1) introduce and practice, and (2) review homework. The Algebra Learning Project modifies this sequence in two ways. First, it recognizes that meaning is constructed gradually and that rarely is the two day contact sufficient for adequate knowledge construction. Therefore, concepts are covered in three to eight days, depending on complexity and importance. Second, since connections between what is known and that which is new constitute a major learning outcome, new concepts are always introduced gradually and in ways which capitalize on pupils' informal or intuitive knowledge.

When integers are introduced, no symbols are used at first. Elevators move up and down; teams move back and forth; businesses have profits and losses. Only after such settings are used to motivate a need and to stimulate some reflections on common experiences are symbols introduced.

From the teacher's perspective, pupils' previous experiences are used to gradually develop ideas over several days. From the pupils' perspectives, previous knowledge is gradually extended and formalized as mathematics is used as a way of describing, recording and reorganizing these experiences and previous knowledge. And from a constructivist perspective, learners are daily constructing new mathematical knowledge which extends the overlap or fit among their knowledge structures through meanings constructed as a result of common educational experiences. Furthermore, the pupil is constructing mathematical knowledge which is an extension of, and therefore more integrated with, previous knowledge and experience.

4.6 Problem Solving

Rachlin *et al.* (1985) use Kantowski's (1980) four levels of problem solving development in their curriculum development work. They are, briefly described,

1. has little or no understanding of strategies and structures;
2. can follow someone else's solution; can suggest strategies previously used; insecure; can participate in a group;
3. can suggest new strategies, can understand and appreciate multiple solution strategies; and
4. can select appropriate strategies; is interested in elegance and efficiency; is interested in novel and challenging problems.

The curriculum, through use of Krutetskii's concepts and nonroutine problems, attempts to move pupils as far through the four levels as possible. From the teacher's and pupil's perspectives, the pupils are becoming problem solvers. From a constructivist perspective, the pupils are developing coping skills which enhance their long-term viability. That is, they are learning how to readily adapt to new environments.

4.7 Reflection and the Role of the Teacher

New topics are introduced through problem solving. Such activities lead pupils to reflect on previous experiences and to extract mathematical relationships from them or to see, articulate and impose mathematics on them. Then throughout the work on the concepts, the teacher pushes (or pulls!) the pupils to see new relationships within and across mathematical knowledge and personal experience. In the hands of a clever teacher, the curriculum leads a pupil to reflect. And as a result of this reflection, pupils relate formal and informal experience, and semantic and syntactic knowledge.

From the teacher's perspective, students are growing as a result of curriculum experiences which induce reflective abstraction. From the pupils' perspectives, they are developing a commanding use of many new relationships. And from a constructivist perspective, the pupils are constructing their own mathematical knowledge.

V. CURRICULAR INTERACTIONS

As before, the objective of this section is to demonstrate the manner in which the key ideas of the project can be presented to achieve dual impact in teacher education and classroom practice. Once again the cycle will be operationalized at two levels. The rationale for doing so is the same that I presented under the conflict teaching.

5.1 Teacher/Pupil Interactions: Krutetskii's Contributions

Reversibility, flexibility and generalization are simultaneously processes and products. As the teacher provokes cognitive conflict through various curriculum activities (process), the pupils engage in metacognitive activity which leads to modifications of their personal realities. At the same time, these experiences help learners disengage from narrow conceptions and misconceptions, to view mathematical operations and relations as reversible, become more flexible in their approaches to problems, and attain a set of rich connections in their mathematical experiences and knowledge structures; these are the products.

5.2 Teacher Educator/Teacher Interactions: Krutetskii's Contributions

As a part of the teacher education experience, teachers are once again challenged in two domains, "how" and "why." I have found the following to be effective:

Week 1: Mini-lecture on reversibility and flexibility; give written examples from other teachers; give copies of Rachlin et al (1985) paper; request examples for the next week;

Week 2: Groups of three go over examples, then whole class sharing; request try-outs with pupils and reaction paper for the next week; mini-lecture on generalization; give written examples from other teachers; request examples for next week;

Week 3: Small group discussions on try-outs and on generalizability ideas and then whole class; repeat week 2 reversibility/flexibility for generalizability;

Week 4: Small groups on generalizability try-outs.

5.3 Examples and Testimonials: Krutetskii's Contributions

As before, the following excerpts are from written work submitted by in-service teachers in graduate courses in 1987. They were written as the reaction/reflection papers described earlier.

5.4 Reversibility and Flexibility

Experiences of Gail Styler, Elementary Teacher

I worked with a five year old, Dan. I chose to explore with him the addition of whole numbers with sums less than or equal to ten.

For flexibility Dan and I used a variety of manipulatives and diagrams. First,

I gave Dan opportunities to construct story problems and to demonstrate their solution with concrete objects. Example: you have four blue shorts and two red shorts. How many do you have altogether? We made a game of going around the house and taking turns making up problems.

The next day, we did a similar activity with Dan drawing the story problems to find solutions. I asked him how many cupcakes we would have altogether if Ellen had three and he had two. He drew the problem and counted.

Another time, I asked Dan to show me how he would find out how many 3 + 4 would be if he used counters. I wrote the sum vertically on a sheet of newsprint. He placed the appropriate number of counters after each numeral. He then placed an equivalent set of counters in a row under the problem. With my prompting, he then wrote a '7' under the problem also.

Dan was unfamiliar with the number line. To introduce it to him, we placed a long piece of tape on our floor and marked it from 0 to 10 using large index cards. I began by explaining that we would use the number line to help us count steps. I directed him to take four steps; then, one step. I asked him how many steps he had taken altogether. He glanced at the number on the floor and gave the correct responses. We made up several more problems and, when a problem called for a sum greater than ten, we discussed alternative methods of showing the answer. His favorite method was to draw them.

For reversibility, I first gave Dan some experience with familiar situations around our home. Example: What do we add to red paint to make it pink paint? What do we add to white paint to make it pink? What color do we get if we mix red and white? What do we add to water to make lemonade? And so on. Next, we tried some arithmetic situations. How many are three plates and four plates? We have three plates; how many do we need to make seven plates? Lastly, using index cards with numerals and symbols, I asked him to find the missing card(s) when I omitted one or two number cards from a number sentence. I allowed Dan to choose any method he wanted to solve his problems. He was adept at figuring the missing addend or sum in his head. He visualized objects to represent the given numerals.

Dan enjoyed the experiences. He particularly likes to draw. However, when drawing is not possible or would take too long, he now has other alternatives for his problem solving.

Experiences of Ruth Siko, Junior High Math Teacher

For this week's activity on flexibility, I worked with two ninth grade boys who are currently enrolled in the first year of a two year Algebra I program. I chose the topic of subtraction of mixed numbers with like denominators with regrouping and used three methods in the activity – Cuisenaire rods, number lines, and diagrams. Because these students should have subtracted mixed numbers for the past several years, I thought it would be interesting to see what they already knew, so I told them to suppose I was a fourth grader who did not know how to subtract with mixed numbers and that I needed to know how much I would have left out of 3 1/4 dollars if I spent 1 3/4 dollars for lunch. I asked one of the boys to be the teacher and show me how to get the answer to 3 1/4 – 1 3/4 by using the Cuisinaire rods. After a few moments, he counted out the correct number of blocks (1 1/2) and said that my change would be one and a half dollars. I reminded him that he must explain more because I didn't understand how he got his answer. He thought longer and after a minute, he said that one of the big blocks would be exchanged for four-fourths and when one of the long blocks, along with three of the shorter blocks were taken away, we would be left with one and a half blocks. I said, "A half?" He pointed to the rods and explained that two-fourths was the same as one half. I gave the other boy a similar problem and asked him to teach me. He went through the mechanics correctly, but as I did with the first boy, I had to remind him to explain more as I was the student and he the teacher. After they did two more problems with the fraction blocks, I told them that we could take the same problems but show the answer by using number lines instead of blocks. I did the problem 1 1/3–2/3 on the number line and then gave each of them a problem that they had done earlier using fraction rods. This time I asked them to demonstrate using the number line.

Finally I demonstrated on the board how to use a diagram with circles and another with rectangles to show subtraction of mixed numbers. Again I asked them to demonstrate to me one of the subtraction problems that each had done using the other ways, and they were able to demonstrate their problems using a diagram of rectangles.

After reflecting on the experience, I will try two changes when I do the activity again. Instead of using a story problem with money to introduce the chosen topic, I think it would have been better to use a more classic fractional story problem, maybe one with cakes or pies. I felt that the money problem didn't clearly focus on the fractional structure of the problem. I would also be careful about doing the diagram method of the chalkboard because the "take-away" parts got erased during the activity, and afterwards I felt that the erasure could present a problem for a beginner by preventing a student from easily identifying the fractional parts.

As far as the boys' reactions, they enjoyed the experience and seemed to be able to relate the three methods to the one topic.

5.5 Generalizability

Experiences of Kathy Atkins, Middle School Math Teacher

I asked an adult, Joe, to define the term average. He had a difficult time putting his thoughts into words. He described the average of a group of numbers as the middle number and inappropriately used the word median in his definition. I then asked him to work the following problem: "The daily temperatures for last week rose to 4°C, 12°C, 14°C, 10°C, 8°C, 6°C, and 16°C. What was the average temperature for the week? He quickly computed the correct answer.

Joe was then instructed to complete the following exercises:

(A) Increase each temperature by 1°. What happens to the average?
(B) Subtract 3° from each temperature. What happens to the average?
(C) What happens to the average when you double each temperature?
(D) What happens to the average when each temperature is divided by 2?

After completing problem A, Joe recognized a pattern and was able to predict correctly the answers to the next three exercises before actually calculating. When I asked him to explain what he had learned from the exercises, he was better able to express his views concerning the concept of averages. He explained that an average is one value that best represents the range of quantities in a set of numbers. "If each number in a group is changed in the same way, then the representation of the group must also change," explained Joe.

I then asked Joe to find the averages of the following groups of test grades:

Group A	99,	100,	98		
Group B	99,	100,	98,	43	
Group C	78,	71,	74,	77	
Group D	78,	71,	74,	77,	100

The results of this exercise did not surprise him. He explained that his experience with averaging grades has taught him that because all of the numbers in a group make equal contributions to the average, one number which is considerably higher or lower than the others can drastically affect the mean.

I wish that I had used this method of instruction when teaching the topic of averages to my sixth graders. When discussing the term a few months ago, my students explained that one computes an average by adding several numbers and then dividing by the number of addends. I was thrilled that they remembered the procedure, and I'm afraid that I did little to expand their perceptions of the concept.

Most students view the concept of averages as merely a series of steps to perform. Teachers often facilitate this idea by centering their explanations around the description of an algorithm rather than around a discussion of the actual meaning of the word "average." Even the textbooks define the concept in terms of the computations to be performed. For example, the sixth grade text defines the term average as the quotient found by dividing a sum by the number of addends. It is no wonder that the students have a narrow conception of the true meaning of the word.

The use of generalization in the classroom can definitely broaden the students' perceptions of the concept. If I had used this instructional method with my sixth graders, I believe that they would have understood that an average is not merely a computed answer but is a description of a group of numbers, a value intermediate to a list of values.

I will definitely consider using generalization when presenting topics in the future. I believe that by using this strategy, I can aid my students in developing more comprehensive conceptions of individual topics as well as of the subject of mathematics in general.

Comments from Darlene McCaskill Math Center Instructor

Generalization is the process of expanding and extending mathematical concepts to encompass a broader range of connections. This should give the students a more interrelated set of mathematical concepts and allow them to use their intuitions to expand their knowledge into new areas. The learners are led to move away from compartmentalized ideas.

In teaching the graphing of parabolas begin with $y = x^2$. Expand this to include $y = 2x^2$, $y = 3x^2$, $y = 4x^2$, then $y = x$, $y = x^2$, $y = x^3$, then $y = x + 1$, $y = x + 2$, $y = x + 3$, and so on. After practice, reflection and discovery, students will have a much richer knowledge base related to parabolas.

5.5 Teacher/Pupil Interactions: Intuition and Time

The advantage of the extended time frame over which content is presented in the Algebra Project materials is that it permits maximum impact on construction *over time*, both cognitively and affectively. When pupils can see what is new as extensions of what is already known in both non-school settings (informal, intuitive) and school settings, they feel the knowledge is theirs; they have created it. Rachlin *et al.* start with informal or intuitive knowledge and extend it over a 3 to 8 day period.

5.6 Teacher Education/Teacher Interactions: Intuition and Time

Teacher educators get teachers to use their intuitions to group students, to design experiences which develop flexibility and to create masterful question sequences which help learners generalize their mathematical knowledge. These activities draw upon some of the most artful aspects of teaching. Also, since a class meets only once each week, teachers are exposed to new ideas over a period of two to four weeks. During that time they report, discuss, read, reflect, react, try-out, and so on.

5.7 Interactions: Problem Solving

In the very process of working with exercises which focus on reversibility, flexibility and generalization, pupils solve problems which push them to find new connections, new ways of viewing information. And in the process of creating new curriculum experiences, and new classroom questioning sequences, teachers strike right at the very core of professional problem solving.

5.8 Interactions: Reflection and Teacher Role

We have seen that the cycle produces pupil reflection and metacognitive activity. We have also seen in the descriptions and teacher comments that the teacher's role has changed considerably from a transmitter of mathematical knowledge to an organizer, planner, facilitator, questioner, helper, monitor.

Similarly, the teacher of teachers has induced considerable reflection among practitioners by forcing them to analyze and reflect about old and new professional practices through participation in small and large group discussions and by writing reaction papers. In the process, the teacher educator has also become much more of a resource person, organizer, and monitor of learning activities.

Collectively, we see learners at both levels, as teachers and as pupils, become much more actively engaged in and responsible for their own learning. Teachers are becoming empowered professionals, and pupils are becoming empowered mathematics learners.

REFERENCES

Bell, A. (1983a). "Diagnostic Teaching of Additive and Multiplicative Problems". In R. Herscovics (ed.), *Proceedings of the Seventh International Conference for the Psychology of Mathematics Education.* (pp. 205–10). Rehovot, Israel: Weizmann Institute of Science.

— (1983b). *Teacher's Handbook.* Nottingham: Shell Centre, U. of Nottingham, UK.

— (1984). "Short and Long Term Learning - Experiments in Diagnostic Teaching Design". In B. Southwell, R. Eyland, M. Cooper, J. Conroy & K. Collis (eds.), *Proceedings of the Eighth International Conference for the Psychology of Mathematics Education.* (pp. 55–62). Darlinghurst: Mathematical Association of South Wales, Australia.

— (1986). "Outcomes of the Diagnostic Teaching Project". In L. Burton & C. Hoyles (eds.), *Proceedings of the Tenth International Conference for the Psychology of Mathematics Education.* (pp. 331–5). London: U. of London.

Kantowski, M.G. (1980). "Some Thoughts on Teaching for Problem Solving". In S. Krulik & R.E. Reys (eds.), *Problem Solving in School Mathematics.* 1980 yearbook (pp. 195–203). Reston, VA: National Council of Teachers of Mathematics.

Kilpatrick, J. (1978). "Research on Mathematical Abilities". Paper presented for the Group in Science and Mathematics Education, Berkeley, CA: University of California.

Krutetskii, V.A. (1969). "An Analysis of the Individual Structure of Mathematical Abilities in School Children". In J. Kilpatrick & I. Wirszup (eds.), *Soviet Studies in the Psychology of Learning and Teaching Mathematics.* (Vol. 2). Stanford: School Mathematics Study Group.

— (1976). *The Psychology of Mathematical Abilities in School Children.* J. Kilpatrick & I. Wirszup (eds.). Chicago: University of Chicago.

Rachlin, S.L., A. Matsumoto & L.A. Wada, (1985a) *Teaching Problem Solving Within the Algebra Curriculum.* Paper presented at the Annual Meeting of the American Research Association, Chicago.

Rachlin, S.L., A. Matsumoto & L.A. Wada, (1985b). *Algebra I: A Process Approach.* Honolulu: University of Hawaii.

Resnick, L. (1981). *The Psychology of Mathematics for Instruction.* Hillsdale, NJ: Erlbaum.

Skemp, R.R. (1979). "Relational Understanding and Instrumental Understanding".

Arithmetic Teacher, 26, 3, 9–15.
Swan, M. (1983). "Teaching Decimal Place Value: A Comparative Study of "Conflict" and "Positive Only" Approaches". In R. Hershkowitz (ed.), *Proceedings of the Seventh International Conference for the Psychology of Mathematics Education*. Rehovot, Israel: Weizmann Institute of Science.
Underhill, R. (1986). "Mathematics Teacher Education: A Constructivist Perspective". Paper presented to the Discussion Group on the Psychology of Training Practicing Teachers of Mathematics at the annual meeting of the International Group for the Psychology of Mathematics Education, London.
von Glasersfeld, E. (1983). "Learning as a Constructive Activity". In J.C. Bergeron & N. Herscovics (eds.), *Proceedings of the Fifth Annual Meeting of the North American Chapter of the International Group for the Psychology of Mathematics Education*. Vol. 1 (pp. 41–69). Montreal: University of Montreal.
Whimbey, A. & J. Lochhead (1986). *Problem Solving and Comprehension* (4th ed.). Hillsdale, NJ: Erlbaum.

Mathematical Education Library

Managing Editor: A.J. Bishop, Cambridge, U.K.

1. H. Freudenthal: *Didactical Phenomenology of Mathematical Structures.* 1983
   ISBN 90-277-1535-1; Pb 90-277-2261-7
2. B. Christiansen, A. G. Howson and M. Otte (eds.): *Perspectives on Mathematics Education.* Papers submitted by Members of the Bacomet Group. 1986. ISBN 90-277-1929-1; Pb 90-277-2118-1
3. A. Treffers: *Three Dimensions.* A Model of Goal and Theory Description in Mathematics Instruction – The Wiskobas Project. 1987 ISBN 90-277-2165-3
4. S. Mellin-Olsen: *The Politics of Mathematics Education.* 1987
   ISBN 90-277-2350-8
5. E. Fischbein: *Intuition in Science and Mathematics.* An Educational Approach. 1987 ISBN 90-277-2506-3
6. A.J. Bishop: *Mathematical Enculturation.* A Cultural Perspective on Mathematics Education. 1988
   ISBN 90-277-2646-9; Pb (1991) 0-7923-1270-8
7. E. von Glasersfeld (ed.): *Radical Constructivism in Mathematics Education.* 1991 ISBN 0-7923-1257-0
8. L. Streefland: *Fractions in Realistic Mathematics Education.* A Paradigm of Developmental Research. 1991 ISBN 0-7923-1282-1
9. H. Freudenthal: *Revisiting Mathematics Education.* China Lectures. 1991
   ISBN 0-7923-1299-6

KLUWER ACADEMIC PUBLISHERS – DORDRECHT / BOSTON / LONDON

GPSR Compliance
The European Union's (EU) General Product Safety Regulation (GPSR) is a set of rules that requires consumer products to be safe and our obligations to ensure this.

If you have any concerns about our products, you can contact us on

ProductSafety@springernature.com

In case Publisher is established outside the EU, the EU authorized representative is:

Springer Nature Customer Service Center GmbH
Europaplatz 3
69115 Heidelberg, Germany

www.ingramcontent.com/pod-product-compliance
Ingram Content Group UK Ltd.
Pitfield, Milton Keynes, MK11 3LW, UK
UKHW020953220926
475831UK00001B/64
9789048140916